MECHANICAL TESTING OF METALLIC MATERIALS
THEORY AND PRACTICE

CHINMAYA MOHAPATRA

SECOND EDITION-2018
Copyright © Mamata Mohapatra

No part of this publication may be stored in a retrieval system, transmitted or reproduced in any way, including, not limited to phyotocopy, photograph without the prior permission of the publisher.

JAY JAGANNATH TECHNICAL PUBLISHERS
718/1, BOMIKHAL, RASULGARH
BHUBANESWAR-751010

Published by:
Mamata Mohapatra for JJTP, Bhubaneswar.

Rs 200/- (Two hundred only)

Printed at:

PREFACE

This book presents elementary concepts of mechanical properties of metallic materials, their significance and how they are affected by crystal structures and defects. Effort has been made to explain various destructive and non-destructive tests carried out on them to evaluate their properties for selection and quality control. Care has been taken to make the book easy going and comprehensive so that it can be used by engineering students of any branch trying to get an idea about perperties of metallic materials and mechanical metallurgy.

It is also important to emphasize here that the diagrams used in the book are highly schematic and may njot be identical to the actual one.

I am indebted to my wife and son for their constant encouragement all along my endeavour to bring out a revised edition of the book first pulished in the year 2009.

My thanks to all those who helped and encouraged me in my academic persuit. I will satisfied if this work is going to help the student to understand this subject better.

Chinmaya Mohapatra

CONTENTS

Chapter 1: **INTRODUCTION TO MECHANICAL TESTING**   1---- 4

Chapter 2: **MECHANICAL PROPERTIES OF METALS**   5-----70

Chapter 3: **TENSILE TEST**   71----105

Chapter 4: **HARDNESS TEST**   106----147

Chapter 5: **FATIGUE TEST**

Chapter 6: **CREEP TEST**

Chapter 7: **IMPACT TEST**

Chapter 8: **ERICHSEN CUPPING TEST**

Chapter 9: **TORSION TEST**

Chapter 10: **BEND TEST**

Chapter 11: **NON DESTRUCTIVE TESTINGS**

Chapter 12: **MAGNETIC PARTICLE INSPECTION**

Chapter 13: **ULTRASONIC FLAW DETECTION**

Chapter 14: **EDDY CURRENT TESTING**

Chapter 15: **X-RAY RADIOGRAPHY**

Chapter 16: **LIQUID DYE PENETRANT INSPECTION**

Chapter 17: **GLOSSARY OF TERMS**

    Bibliography:

Chapter 1
INTRODUCTION TO MECHANICAL TESTING

1.1. Introduction:

With rapidly expanding industrial growth and our foray into space, many types of new materials along with the existing materials have been developed to work satisfactorily under specific condition. However, the suitability of the material that it would perform under the given set of conditions has to be initially evaluated in the laboratory with extreme preciseness depending on the criticality of the component. Hence, it has become extremely important to evaluate the properties of the material in a laboratory those are important to assess its life under the actual working conditions. Mechanical testings in general help us to choose a particular material from among different kinds which would be the most suitable under a given set conditions. Mechanical testings are also conducted routinely on the materials which have long been established as a standard material for the purpose to ascertain their quality. For any particular material, the average values of its mechanical properties are listed in standard handbooks which may of course vary widely. Wide variance in properties may be due to change in the following parameters:

i. Source of the raw material.
ii. Method of manufacturing.
iii. Care taken during manufacturing.
iv. Presence of impurities or other defects in the material.

Finally, the purposes of testing the materials are to ensure:

i. Required properties in the defined range.
ii. Quality of the material to serve a purpose.
iii. A basis for comparison between different materials.
iv. An integral part of research for developing new materials.

1.2. Classification of Properties:

Properties of engineering materials may be classified broadly into three groups such as:

    i. Physical
    ii. Chemical
    iii. Mechanical

1.2.1. Physical Properties:

These are the bulk properties of the material like density, specific gravity, porosity, moisture content and etc.

1.2.2. Chemical Properties:

These are properties related to the chemical reactivity of the materials. These properties are important while studying about corrosion, oxidation or reduction characteristics of the materials.

1.2.3. Mechanical Properties:

It is important to evaluate certain bulk properties such as strength, hardness and etc of the materials for engineering purposes. The tests conducted to evaluate such poperties are usually known as mechanical tests. Depending on the nature the tests may be classified as:

    i. Destructive test.
    ii. Non - Destructive test.

1.3. Destructive Tests:

These test are termed so because during these tests the specimen or materials are fractured or deformed and cannot be used further or rather destroyed.

As different properties are to be evaluated for different applications, various tests are conducted on the materials when required. Common destructive tests are:

i. Tensile
ii. Compressive
iii. Hardness
iv. Fatigue
v. Creep
vii. Impact or Fracture

1.4. Non-Destructive Tests:

It is to emphasise here that during the nondestructive tests the specimen or material is not deformed or broken and it can be used after the test if found suitable. Various non-destructive tests employed to determine the quality of the materials are:

i. Visual
ii. Leakage
iii. Magnetic Particle
iv. Dye Penetrant
v. Acoustic Or Ultrasonic
vi. Eddy Current
vii. X-ray Radiography.

1.5. Test Types:

To evaluate various mechanical properties the materials may undergo two types of tests as listed below:

i. Standard Laboratory tests.
ii. Service Condition tests.

i. Standard Laboratory Test:

When the test is carried out under pre-specified standard conditions mostly having the variables at their ideal places it is termed as standard laboratory tests.

ii. Service Condition Test:

When the test is carried out under the actual service conditions it is called service condition test.

1.6. Tests on Raw Material & Finished Products:

Of the above listed two test practices, the in-house standard laboratory tests are indispensable in evaluating the mechanical properties of the material to assess its suitability for a particular purpose. Let us take the case of a particular grade of steel used to make gears for an automobile plant. From our previous experience purchase order for a particular grade of steel is placed with the steel plant. The raw material (a particular grade of steel) received from the plant undergoes number of standard tests to ascertain its quality for which the order was placed. After complete manufacturing of the component (gear) further tests are carried out on it to certify its suitability for the final application. However, the tests carried out on raw- materials & finished products may vary widely.

****

Chapter 2
MECHANICAL PROPERTIES OF MATERIALS

2.1. Introduction:

Metallic materials are frequently chosen for structural applications as they have the desirable combination of mechanical & physical properties. Hence, our present discussion is confined only to mechanical behaviour of metals & alloys.

2.2. Concept of Stress & Strain:

Let us consider a uniform cylindrical bar being subjected to an axial load as shown in the fig.2.1. The change in its length over the original length L_0 is defined as Strain (e).

Mathematically, strain $e = \dfrac{\delta l}{L_0} = \dfrac{\Delta l}{L_0} = \dfrac{L - L_0}{L_0}$.

The factor e is defined as *average linear strain* or simply the *strain*. It is obvious that strain has no unit or dimension simply a numerical value. As shown in the free body diagram the external load P is balanced by the internal resisting force: $\int \sigma \times dA$, where factor σ is defined as the stress distributed uniformly over the area A.

Now, $P = \int \sigma \times dA$ or, $P = (\sigma \times A)$.

Hence, $\sigma = P/A$, is called the stress. The stress is expressed in Kg-f or Newton/mm².

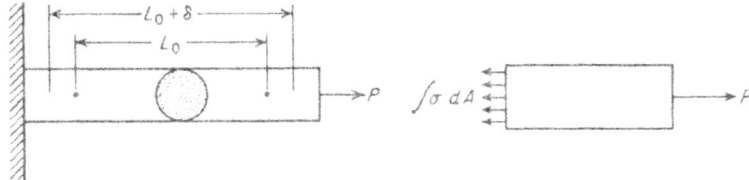

Fig.2.1. Cylidrical Bar Subjected to an Axial Load with its Free Body Diagram.

2.3. Elastic-Plastic Nature of Materials & Young's Modulus:

Solid materials can be deformed when subjected to an external load. It is observed that upto a certain limiting load, the deformed solid will regain its original shape & size once the load is removed. The deformation which is recovered on removing the load is termed as elastic deformation. The limiting load beyond which the material no longer behaves elastically is known as elastic limit. When the elastic limit of the material is exceeded, the body undergoes permanent deformation and never regains its original shape & size even if the load is removed. Permanent deformation is known as plastic deformation. Both the deformations are shown schematically in the fig.2.2.

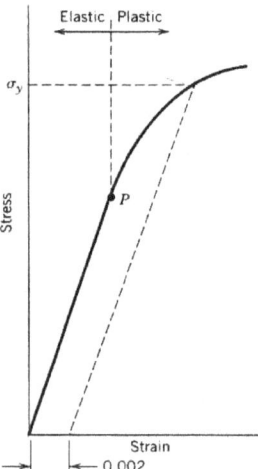

Fig.2.2. Elastic - Plastic Deformation and Elastic Limit

From the figure it is important to note that within the elastic limit the stress is directly proportional to strain produced in the metallic materials. The mathematical formula that relates both stress and strain wihin the elastic limit is known as Hooke's law.

Hooke's law is repsented as $\sigma = E \times e$, where, σ, e & E are the stress, elastic strain and Young's modulus or Modulus of elasticity respectively. Hooke's law is a valid relationship for engineering designs for metallic materials. The straight line portion of the stress-strain digram is shown schematically in the figure 2.3 and its slope is termed as Young's or Elastic modulus.

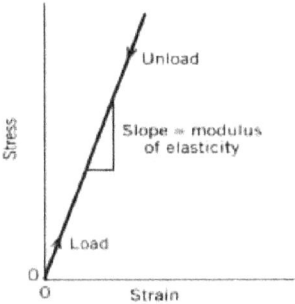

Fig.2.3. Straight line portion of the Stress- Strain diagram.

Young's modulus E is one of the *four elastic* constants of the material and is as an *intrinsic* property of the material. The other elastic constants of an isotropic material are Shear Modulus G, Bulk Modulus K & Poisson's Ratio μ. All the elastic constants of the metallic material is influenced by the following factors:

i. Type of bond(s) that exist in the material.

ii. Temperature at which the constants are measured.

iii. Anisotropy of the material.

2. 4. Shear Modulus:

The ratio of shear stress τ to the shear strain γ within the elastic limit is defined as the Shear modulus or modulus of rigidity G. This ratio can be written mathematically as, $\tau = G\gamma$.

Further, it can also be proved that $G = E/2(1+\mu)$.

2.5. Bulk Modulus:

A material under three dimensional stress condition is subjected to σ_x, σ_y and σ_z along X,Y & Z axes respectively. Let the initial volume of the material V_0 changes by dV due to application of three dimensional stress. So the volume strain, $e_v = dV/V_0$ and bulk modulus, $K = \bar{\sigma}/e_v = E/3(1-2\mu)$

Poisson's ratio μ is 0.25 for a perfectly isotropic elastic material. For most metals, the values of μ is close to 0.33 which can be proved mathematically. For ideal plastic materials (elastomers) the value of μ is 0.50. For metals and alloys the value of, μ usually ranges from 0.25 - 0.35. For anisotropic materials like composites the value of μ even may exceed 0.50.

2. 6. Poisson's Ratio:

When a material is subjected to tensile stress in a particular direction X, it elongates in that direction while it contracts in the other orthogonal directions i.e. Y & Z as shown in the fig. 2.4.

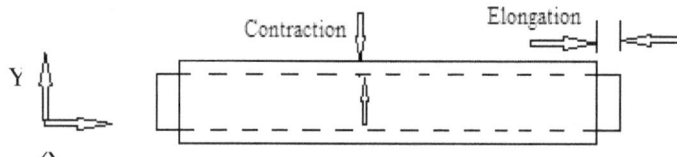

Fig.2.4. Mateial showing Elogation in the X-axis and Contraction in the Y-axis when subjected to Tensile Stress.

The transverse strains, e_y or e_z always bear a constant ratio with the longitudinal strain e_x. The ratio of transverse strain to longitudinal strain ($e_y : e_x$ or $e_z : e_x$) is known as Poisson's Ratio.

This ratio can be mathematically written as: $\mu = e_y/e_x = e_z/e_x$ or, $e_y = e_z = -\mu e_x$. The negative sign in the formula signifies contraction in the Y- axis or direction.

2.7. Isotropy, Anisotropy & Orthotropy:
I. Isotropy:

The material which exhibits identical properties in all directions of the matrix is known as an *isotropic material* and such a phenomenon is known as *isotropy*. Isotropic materials obey Hooke's law. Usually metals & alloys are considered to be isotropic.

II. Anisotropy:

Materials exhibiting dissimilar properties in different directions of the matrix are known as anisotropic materials and such a phenomenon is known as *anisotropy* or *anisotropism*. Quartz crystal and composites are considered to be anisotropic materials.

III. Orthotropy:

Wood is an example of an orthotropic material. Material properties in three perpendicular directions (axial, radial, and circumferential) are different. In particular, the mechanical properties (such as strength and stiffness) along the grains are typically larger than in the radial and circumferential directions. Another example of an orthotropic material is a metal which has been rolled to form a sheet. The properties of the rolled sheet in the rolling direction and other two transverse directions will be different due to the anisotropic structure that develops during rolling. Orthotropic materials are a subset of anisotropic materials.

2.8. Stress – Strain Diagram:

The relation between stress & strain is generally shown by plotting a stress versus strain diagram which is called an engineering stress-strain diagram. This diagram is useful to design engineers in calculating other important mechanical properties of the materials essential in designing. The general stress-strain diagrams of brittle and ductile materials are shown relatively in the fig.2.5.

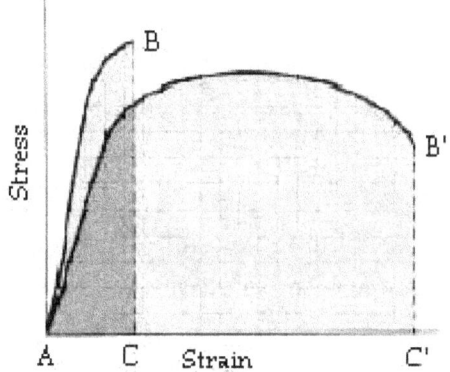

Fig. 2.5. AB & AB' Curves Represent Brittle and Ductile Materials respectively.

2.9. Yielding and Yield Strength:

Most of the engineering structures are designed only to withstand elastic deformation. It is therefore, desirable to know the stress level at which permanent deformation starts so that a lower load may be applied on the material so as to limit the deformation within the elastic range.

The onset of plastic deformation in the material is better known as yielding. The stress at which yielding or plastic deformation starts is termed as **yield strength**. As a matter of fact yield point always lies above the elastic limit of the material.

2.10. Ductility:

Ductility is a mechanical property which measures the extent of plastic deformation sustained by the material at fracture. The material which exhibits large plastic deformation prior to fracture is known as ductile material while a material which exhibit very little or nil plastic deformation prior to fracture is known as brittle material. Ductility may be expressed quantitatively either as percentage elongation or as percentage reduction in area at fractute over the initial value. Percentage(%) elongation $= L_f - L_o / L_o \times 100$ & % RA $= (A_f - A_o)/A_o \times 100$ where L_o & L_f are the initial & final gauge lengths, A_0 & A_f are the initial and final cross-sectional areas of the specimen at failure respectively.

2.11. Resilience:

The ability of a material to absorb energy when deformed elastically and release the same when unloaded is called resilience. Resilience is usually measured by a quantity called modulus of resilience, U_R. This is defined as the elastic strain energy per unit volume required to stress the material from totally stress free condition to yield stress. Under uniaxial tensile stress condition the elastic strain energy per unit volume is: $U_R = \sigma_0 e_0 / 2 = (\sigma_0 / E) \times \sigma_0 / 2$

As per definition modulus of resilience is: $U_R = \sigma_x e_x / 2$

where σ_0 the yield stress or elastic limit, e_0 is the strain at the elastic limit and E is the modulus of elasticity. Considering the stress-strain diagram for general metals as shown in the fig.2.6. the area under the curve upto the elastic limit (the hatched area in both the cases) is the quantitative measure of the resilience of that material.

Thus, resilient materials have high yield strengths and low modulus of elasticity. Such materials are extensively used to manufacture springs.

2.13. Toughness:

The ability of engineering components to withstand occasional stress above yield stress within the plastic range is extremely important to resist fracture. This ability is generally defined as toughness of the material. Quantitatively toughness is the area under the stress-strain curve including the hatched portion upto fracture as shown in the figure2.6 for both brittle & ductile materials. Generally, materials of higher ductility are tougher than materials with lower ductility. The materials which are not tough are generally brittle. Hence a structural steel is tougher than spring steel. Toughness enables materials to withstand occasional impact load above its yield strength without causing failure. Toughness is the ability of a material to absorb energy in plastic range upto fracture.

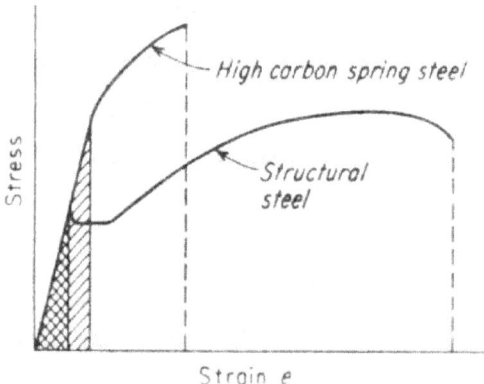

Fig. 2.6. Schematic representation of Resiliance and Toughness.

Mathematically toughness can be expressed as:

I. For Ductile Mateials: $U_T = (\sigma_0 + \sigma_u/2) \times e_f$ -----------(2.1)

II. for Brittle Matrials: $U_T = 2\sigma_u \times e_f / 3$ ----------(2.2)

where, σ_0 is the yield strength, σ_f is the ultimate strength, e_f is the strain at failure and U_T is the toughness of the material under test.

2.14. True Stress and True Strain:

The engineering stress–strain diagram does not give a true indication of the deformation characteristics of ductile materials as it is based entirely on the original dimensions of the specimen which change continuously during the test. The engineering stress–strain diagram shows a decline in the stress level on continued deformation beyond ultimate limit (σ_u) indicating as if the material is becoming weaker with continued deformation.. But this fact is not true. As a matter of fact, the strength increases up to fracture. At the onset of necking beyond the ultimate limit, the area of cross–section resisting the applied load decreases rapidly reducing the load bearing capacity of the material. So it is more meaningful to use the true stress-strain diagram than the engineering stress-strain diagram for all design purpose. Both the curves are shown schematically in the fig.2.7.

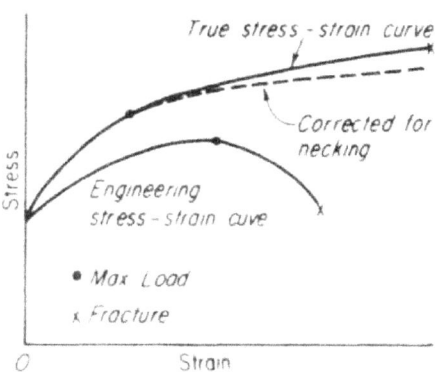

Fig.2.7. Schematic True and Engg. Stress-Strain Diagrams.

2.15. Relation between True and Engineering Stress:

True stress is defined as the load P upon the instantaneous cross–sectional area A_i of the deforming specimen. Mathematically,

True stress $\sigma_T = P/A_i$ & Engineering stress $\sigma = P/A_o$

Hence, we can write: $\sigma_T = P/A_i = (P/A_o) \times (A_o/A_i) = \sigma(A_o/A_i)$

A_0 is the initial cross-sectional area of the specimen that resists the load. A_i is the instantaneous cross-sectional area of the specimen during the test. Applying the principle of constancy of volume we have: $A_o L_o = A_i L_i$. From this relation it can be deduced that:

$$A_0/A_i = L_i/L_0 = (L_0 + \Delta L)/L_0 = 1 + \Delta L/L_0 = (1+e)$$

Hence, $\sigma_T = \sigma(A_o/A_i) = \sigma(1+e)$

where, σ is the Engineering stress, σ_T is the True stress,

L_0 is the original gauge length,

L_i is the instantaneous length of the sample at any instant,

ΔL is the change in length and

$e = \Delta L/L_o$ is the engineering strain.

2.16. Relation between True and Engineering Strain:

True strain, ε is defined as the change in length over the instantaneous length (L_i) while the engineering strain, e is defined as change in length over original gauge length(L_0).

Mathematically, $e = \dfrac{\Delta l}{L} = \int_{L_0}^{L} dL$.

This relation for engineering strain is valid only within the elastic limit where the strain is rather very small. However, during plastic deformation as the strain is much larger, instantaneous lengths must be considered for calculation of true strain at every instant.

Hence, true strain: $\varepsilon = \sum L_1/L_2 + L_2/L_3 + --$ (as per Ludwig)

Or, $\varepsilon = \int_{L_0}^{L} \dfrac{dL}{L} = \ln(L/L_0) = \ln[(L_0 + \Delta L)/L_0] = \ln[1 + \Delta L/L_0]$

True strain, $\varepsilon = \ln(1+e)$ where e is the engineering strain.

2.17. The Flow Curves:

The true stress-strain curve of a material obtained under uniaxial tensile test condition is known as flow curve. The flow curve indicates the stress required to cause the plastic flow of a material to any given strain. The flow curve is commonly represented by a power expression of the form:

$\sigma_T = K\varepsilon^n$ where, K is the stress at $\varepsilon = 1.0$ and n is the strain hardening coefficient which varies from material to material. The value of n is less than 1.0 for general metals.

This equation is valid form the beginning of plastic flow to the maximum load at which the specimen begins to neck down. On the basis of the above equation, idealized flow curves for different materials as shown in the fig.2.8.

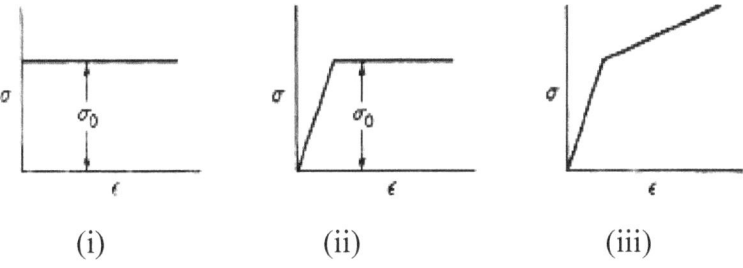

(i)　　　　　　　(ii)　　　　　　　(iii)

Fig.2.8. Flow Curves: i. Rigid Ideal Plastic Material.
　　　　　　　ii. Ideal Plastic with Elastic Region.
　　　　　　　iii. Strain Hardening Material.

2.19. Concept of Plastic Deformation:

Permanent deformation of a material due to application of load is known as plastic deformation. The ability of the metals & alloys to undergo plastic deformation is probably their most outstanding characteristic in comparison with other materials. Different shaping operations such as rolling, extrusion, forging, drawing & etc. involves plastic deformation. The mechanism of plastic deformation could be studied in depth after the discovery of X-ray diffraction by Von Laue in the year 1912.

Fundamentally, it is to be understood that metals and alloys are composed of atoms arranged in specific geometric patterns called lattice and their plastic deformation behaviour is related to their lattice structures. Much of the fundamental work on plastic deformation of metals & alloys has been performed on single crystals with a view to eliminate the complicating effects of grain boundaries & the restraining effects imposed by the neighbouring grains surrounding a particular grain. The concept of plastic deformation of single crystals was further expanded to understand the plastic deformation of polycrystals. After several investigations it has been concluded with certainity that plastic deformation of metal and alloys is either due to slip or twinning or a combination of both.

2.20. Deformation by Slip:

The usual method of plastic deformation in metals alloys is due to sliding of a block of crystal over another along definite directions on definite planes. The planes on which such sliding takes place is known as slip plane and the direction along which the sliding takes place is called slip direction.

This slip phenomenon is very much analogous to the distortion provided to a pack of cards pushed from one end. Slip will occur if the applied shear stress on a plane exceeds a critical value. A step will be produced due to slip on the polished surface as shown in the fig.2.9. On observation under microscope the slip produced will appear as lines called slip lines. The fact, that a single crystal remains a single crystal, even after homogeneous plastic deformation imposes limitations on the way by which plastic deformation may occur. Investigations have clearly shown that slip occurs most readily on specific crystallographic planes in specific crystallographic directions called slip planes and slip directions respectively.

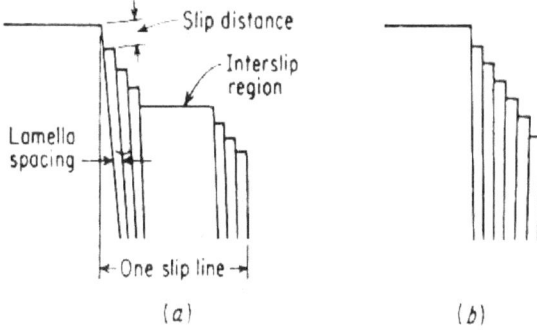

Fig.2.9. Slip in the Material

2.21. Slip Plane & Slip Direction:

Generally slip plane is the plane of greatest atomic density and slip direction is the closest-packed direction on the slip plane. Since the planes of greatest atomic density are the most widely spaced planes in the crystal structure, the resistance to slip is generally lower for these planes than for any other set of planes.

The slip-plane together with the slip direction constitutes the **slip system** of the crystal. In a particular crystal the total number slip systems available for plastic deformation is: the product of: number of slip planes & number of slip directions. Hence for different crystal structures different numbers of slip systems are available. The slip systems available in a crystal structure reflect the ease with which plastic deformation takes place in that crystal structure. Higher the number of slip systems available easier will be its defoemation and higher will be its ductility & malleability.

2.22. Slip Systems for Different Crystal Geometry:
I. HCP Materials: Be, Zn, Cd, Mg, Co, Zr &etc.
i.No. of Slip Planes: The only closest packed plane available in hcp for slip is the (0001) plane better known as the basal plane.

ii.No. of Slip Directions: $3, <11\bar{2}0>$ family of directions. There are only three equivalent directions in this family of directions.
iii. Total No. of Slip Systems In HCP: $1 \times 3 = 3$
(Number of slip planes $\times$ Number of slip directions)

The directions and the plane are shown in the fig.2.10.

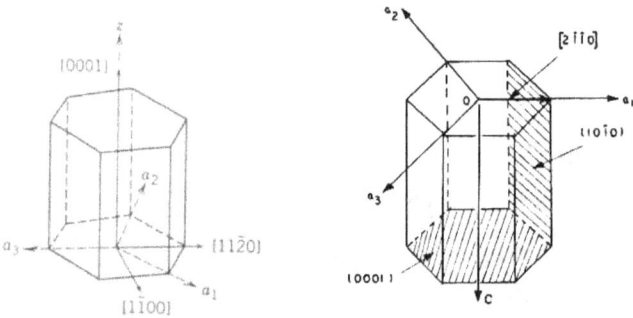

Fig. 2.10. Planes and Direction of Slip in HCP Crystal Structure.

II. FCC Materials:

Ni, Cu, Ag, Pt, Au, Al, Pb, Fe (above $910^0 C$) &etc.

i. No. of Slip planes: Family of (111) planes = 4 sets.

ii. No. of Slip directions: Family of <110> directions = 3 sets.

iii. Total no of Slip Systems: $4 \times 3 = 12$

The slip planes and slip directions are shown in the fig.2.11.

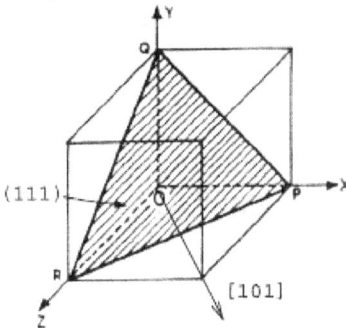

Fig.2.11. Family of Slip Planes & Directions in FCC Structure

III. BCC Materials:

Fe (below $910^0 C$),Cr, Mo, Nb, Ta, V and W.

i. No. of Slip planes: Nil

ii. No. of Slip directions: [111] family of directions

iii. The total no.of slip systems at ordinary temperature: Nil (0).

There is no slip plane at ordinary temperature on which slip can take place. However, (110),(112)&(123) family of planes exhibit slip at elevated temperatures in bcc crystals.

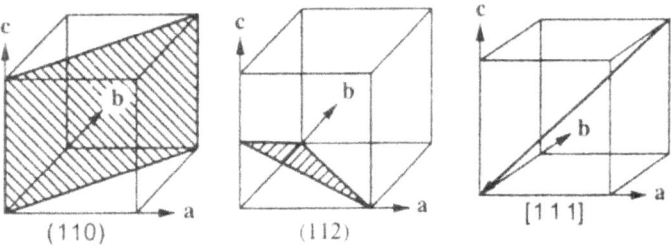

Fig. 2.12. Slip Planes and Direction in BCC Crystal Structure.

However, number of slip systems at elevated temperatures becomes $16 \times 3 = 48$. For this reason bcc materials exhibit much inferior ductility compared to fcc/hcp at lower temperatures while at elevated temperatures they exhibit much higher ductility because as high as 48 slip systems become operational in them.

2.23. Theoretical Shear Strength of a Perfect Lattice:

When slip is assumed to be the translation of one plane of atoms over another then it is possible to make a reasonable estimate of the shear stress required for such a movement in a perfect lattice. Let us consider two planes of atoms subjected to a homogeneous shear stress as shown in the fig.2.13 with the following assumptions:

i. The shear stress is assumed to be acting on the slip plane in the slip direction of magnitude τ.

ii. The atoms are being separated by a lattice distance, a due to the action of shear stress.

iii. The interatomic distance in the slip plane is, b.

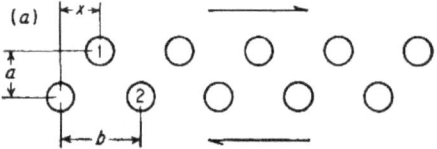

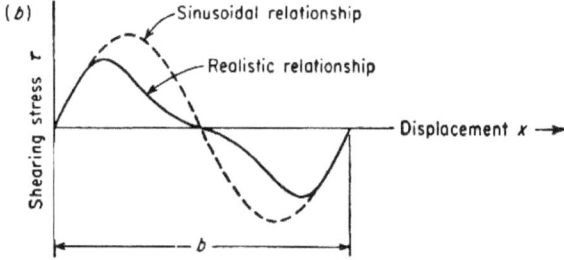

Fig. 2.13. Variation of Shearing Stress during slip between two consecutive atomic planes.

iv. The shear stress causes a slip or displacement in the slip direction of magnitude x.

v. Further, the shear stress in zero initially when two planes are in coincidence. Similarly the shear stress is also zero when the two planes have moved identity distance b because of the symmetry position. The shear stress is also zero when the atoms on the top plane are midway between those of the bottom plane.

As a first approximation the relation between the shear stress and displacement can be expressed by a sine function(ref.fig.2.13).

Mathematically, $\tau = \tau_m \sin(2\pi x/b)$ -----------(1)

Shear stress, τ_m and b are the amplitude & period of the sine wave respectively. Within the elastic limit that is for small values of displacement x as Hooke's law is applicable we can deduce:

$\tau = G \times \gamma = G \times (x/a)$ ---------(2).

Here, G is the shear modulus and γ is the shear strain.

Furthere shear strain, $\gamma = \dfrac{\text{Lateral Displacement}}{\text{Transverse Distance}} = \dfrac{x}{a}$

Now, for small values of θ, $\sin\theta = \theta$ --------(3)

Hence, $\sin(2\pi x/b) = 2\pi x/b$

So, we have, $\tau = \tau_m(2\pi x/b)$ ------------(4)

As a & b are atomic parameters, we can write $a \cong b$

Therefore, we can write, $x/a = x/b$ --------(5)

Combining eq. (2) & (4), we have: $\tau = G(x/a) = \tau_m \times 2\pi(x/b)$

Or, $\tau_m = G/2\pi$ which is the theoretical shear strength of the material.

As shear modulus of the metals is in the range of 20-150 Gpa, their theoretical strengths should lie the range of 3-30Gpa. However, the actual shear stress required to initiate plastic deformation is in the range of 0.5-10Mpa only. This brings up a huge discrepancy between the theoretical & practical strengths of the metallic crystals. The actual strength of a real metallic crystal is in the range of $1/100 - 1/1000^{th}$ of its theoretical strength. This large difference between theoretical & actual strength of a metallic crystal has compelled us to look for other mechanisms of plastic deformation rather than shearing of planes of atoms over each other.

2.24. Slip by Dislocation Movement:

The concept of dislocation was introduced to explain the discrepancy between observed & theoretical shear strength of metals. For this concept to be valid it is necessary to show:

i. That the motion of a dislocation through a crystal lattice requires a stress smaller than the theoretical shear stress.

ii. That the movement of the dislocation in the crystal produces a step of slip band at the free surface as shown in the figure 2.14.

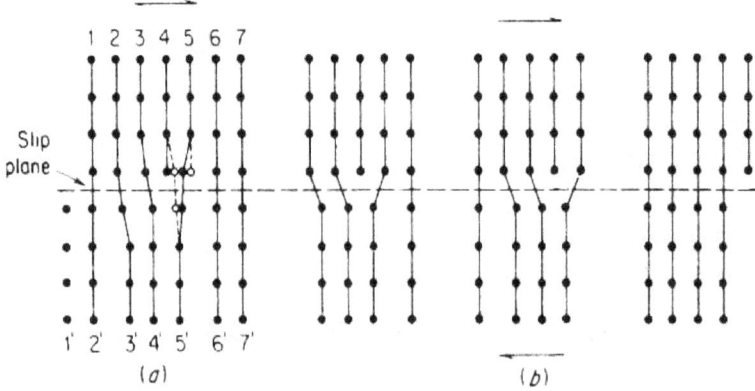

Fig.2.14. Dislocation Movement and Slip.

Considering the figure which ultimately shows slip, it can be seen that the extra half plane of atoms is at the position 4 initially. Under the continued action of the shear stress, a very small atomic movement to the right allows this extra half plane of atoms to line up with 5' simultaneously creating a new half plane of atoms at 5 from its neighbours above the slip plane. By this process the extra half plane of atoms move from its initial position between planes 4' and 5' to a new position between planes 5' and 6'. This movement of extra half plane of atoms from $5 \to 6 \to 7$ will eventually cause the half plane of atoms to come out of the crystal and results in a slip-step in the upper half the crystal on the slip plane.

The existence of a half plane of atoms either in the upper or lower half of the crystal is considered to be a defect in the lattice and termed as dislocation. Dislocation of this nature which produces a step on the free surface of the crystal is termed as edge dislocation. This process of dislocation flow is associated with an energy barrier, ΔE. Hence, it is logical to assume that rather than all the atoms moving simultaneous along the slip plane during slip, only the atoms on the dislocation line and few atoms surrounding the dislocation line make simultaneous movement to facilitate the movement of dislocation line through the lattice in discrete steps. Such a mechanism minimizes the energy required for slip considerably. Slipped material will grow at the expense of unslipped material steadily in the crystal which can be observed physically. To minimize the energy for transition, we make the width of the dislocation or interface thickness (w) to be narrow. **Peierls stress** is the shear stress required to move a dislocation through a crystal lattice in a particular direction.

Mathematically,

$$\tau_p \approx \frac{2G}{1-\mu} \times e^{-2\pi w/b} \approx \frac{2G}{1-\mu} e^{-[2\pi a/(1-\mu)b]}$$

Where, a is the distance between slip planes and, b is the distance between atoms in the slip direction, μ is the Poisson's ratio of the material, w is the width of the dislocation, G is the shear modulus of the material and, τ_p is the *Peierls stress*.

2.25. Critical Resolved Shear Stress (CRSS) for Slip:

Slip in a single crystal depends on the following factors:
i. Magnitude of the shearing stress produced by the external loads.
ii. Geometry of the crystal structure.
iii. Orientation of the slip planes with respect to the shearing stresses.

Slip begins when the shearing stress on the slip plane in the slip direction reaches a threshold or critical value called Critical Resolved Shear Stress(CRSS). The value of CRSS is also influenced by the composition and the temperature at which the test is being conducted. This value is the single crystal equivalent of the yield stress of polycrystalline materials. The fact that tensile loads of different magnitude are required to produce equivalent slip in a single crystal of different orientations can be explained in terms of CRSS. This was first recognised by Schmid. However, it is important to know the orientation of the slip plane with respect to the tensile axis within the single crystal by X-ray diffration prior to determination of CRSS.

2.26. Schmid's Law:

Let us consider a cylindrical single crystal as shown in the figure.2.15 with cross-sectional area of A. The slip plane is inclind to the load axis. Angle between the normal to the slip plane & tensile axes is ϕ and the angle between tensile axis and slip direction is λ.

Now we have:

i. The area of the plane at an angle ϕ to the normal = $A/\cos\phi$.

ii. The component of the axial load P in the direction of the slip on the slip plane. Therefore,

$$\text{CRSS}, \tau_R = \frac{P\cos\lambda}{A/\cos\phi} = \frac{P}{A}\cos\lambda\cos\phi = \sigma\cos\lambda\cos\phi.$$

iii. The shear stress produced on the slip plane is maximum when $\phi = \lambda = 45°$ is: $\tau_R = \frac{P}{2A} = \frac{1}{2}\sigma$, where, σ is the tensile stress.

iv. If the tensile axis is normal or parallel to the slip plane, where either $\lambda = 0°$ & $\phi = 0°$ or $\phi = 90°$ & $\lambda = 0°$ CRSS will be *zero* as $(\lambda + \phi) = 90°$. Slip will not occur as no shear stress develops on the slip plane for these extreme orientations. Crystal close to these orientations tends to fracture or twin. Thus it is not possible to produce slip on a given plane when the plane is either parallel or perpendicular to the tensile stress axis. Further, the Critical Resolved Shear Stress is also dependent on alloying and temperature.

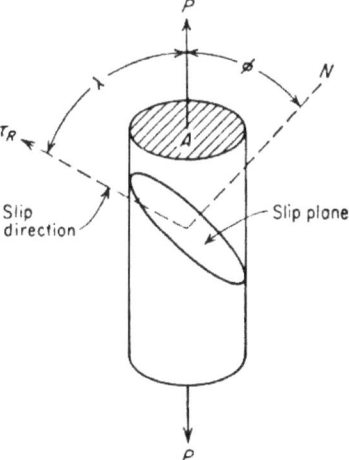

Fig.2.15. Diagram for Calculating Critical Resolved Shear Stress.

Addition of small amount of alloying elements always increases the CRSS. The CRSS decreases as the density of crystal defects in the lattice decreases provided that the total number of defects is not zero. When the last dislocation is eliminated, the CRSS increases abruptly to the theoritical shear strength of a perfect lattice. The ratio of resolved shear stress to the axial stress is called **Schmid's factor, *m***. For a single crystal, $m = \cos\phi.\cos\lambda$.

A single crystal will slip when the resolved shear stress on the slip plane reaches a critical value. This is known as Schmid's law & best demonstrated by the hcp metals.

2.27. Deformation by Twinning:

The second most important mechanism by which metals and alloys deform is the process of twinning. During slip steps are created on the free surfaces of the crystal but the orientation of the entire crystal remains unchanged even after the slip. The phenomenon of twining is shown schmatically in the fig.2.16.

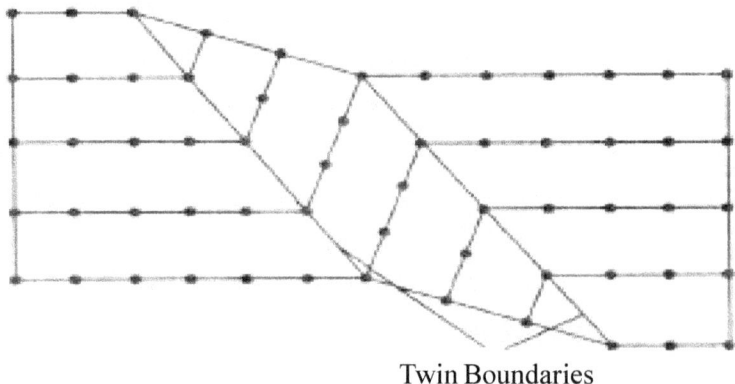

Twin Boundaries

Fig.2.16. Schematic drawing of a Twin.

However, during twinning the orientation of a portion of lattice changes with respect to rest of the lattice. Twinning results when a portion of the crystal takes up an orientation that is related to the orientation of the untwinned portion in a definite symmetrical way. The twinned portion of the crystal is a mirror image of the parent lattice. The planes of symmetry which separate the deformed zone from the undeformed portion of the parent lattice are called the twin planes. Twin planes always occur in pairs. The volume of material which has the mirror image orientation of the matrix material is called the ***twinned zone.*** Twins may come into existence during crystal growth or may be formed during severe plastic deformation.

2.28. Classification of Twins:

There are two kinds of twins which are of interest to the metallurgists. They are:

i. Deformation or Mechanical twins

ii. Annealing twins:

i. Deformation or Mechanical Twins:

This is formed during plastic deformation and most prevalent in hcp metals such as magnesium, zinc and bcc metals like iron and tungsten.

ii. Annealing Twins:

This is most prevalent in fcc metals. These metals which have been cold worked previously twin during annealing due to change in normal growth mechanism.

2.29. Differences between Slip and Twin:

I. Slip:

i. Crystal orientation above and below slip plane remains identical during slip and the original lattice orientation is retained.

ii. As slip usually occurs in discrete multiple of atomic spacing. Hence this mehanism enables large plastic deformation.

iii. Slip occurs on relatively widely spaced crystal planes. They appear as thin lines under microscope.

iv. Slip is the most important mechanism of deformation at ambient and elevated temperature.

v. Slip requires fairly long time for its initiation.

II. Twining:

i. In twinning there is a change in the crystal orientation across the twin boundaries. Hence lattice orientation changes in the twinned zone.

ii. In twinning atomic movements are highly restricted and always less than an atomic distance. This makes the plastic deformation highly localized. The extent of deformation is also small which depends on the distance between the twin planes.

iii. In the twinned zone of a crystal every atomic plane is involved in deformation and they appear as broad line or bands under microscope.

iv. Twinning is an important mechanism of deformation at low temperature or sub-zero temperatures.

v. Twin can be produced in microsecond time.

2.29. Dislocation Theory:

A dislocation is a linear or line defect around which some of the lattice atoms are misaligned. Depending on the geometry of the line defect dislocation can be defined as:

i. Edge dislocation and
ii. Screw dislocation.

2.29.1. Edge Dislocation:

If extra half plane of atoms terminates within the crystal lattice, the defect produced is termed as edge dislocation.

It is a linear defect that centers on a line which is defined along the end of the extra half plane of atoms. Edge dislocation is further classified as positive & negative edge dislocation depending on the location of the extra half plane of atoms with respect to the slip plane. When the extra half plane of atoms appears above the slip plane the edge dislocation produced is termed as positive and if the extra half plane of atoms appers below the slip plane the dislocation is known as negative edge dislocation. Schematically they are represented as (⊥) & (⊤ respectivly in the fig 2.17.

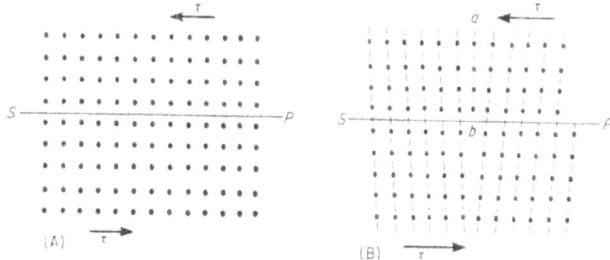

Fig.2.17. (A) Perfect lattice . (B) Lattice model with an Edge Dislocation.

2.29.2. Screw Dislocation:

Contrary to edge dislocation, there is no extra half plane of atoms in case of screw dislocation. Screw dislocation forms when a part of the crystal displaces angularly over the remaining part. The plane of atoms converts into a helical surface or a screw along the interface as shown schematically in the fig.2.18. The angular displacement is similar to movement of a screw when turned. The name screw is derived from this feature.

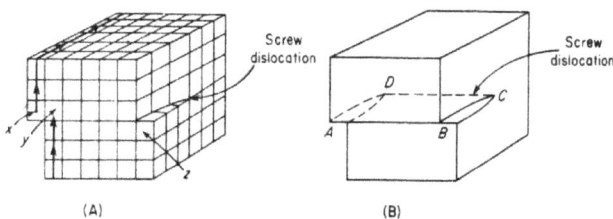

Fig.2.18. Screw Dislocation.

It is symbolically represented and referred to as clockwise and anticlockwise screw dislocation. Finally the dislocation line can be seen as the boundary between the slipped and unslipped parts of the crystal. Therefore, the dislocation is a discontinuity at which the lattice shifts from unsheared to sheared state.

2.30. Burgers Vector & Burgers Circuit:

The magnitude & direction of a dislocation can be determined by a vector called **Burgers vector** named after its originator. It is normally denoted by B.V.or $\bar{b}$. Burgers vector of a dislocation is found by drawing a circuit around the dislocation line called burgers circuit as shown in the fig.2.19. This circuit is analogous to the electrical circuit which encircles the the dislocation line completely by covering equivalent distances in all directions. Burgers vector is the vector which closes the burgers circuit completely. The line drawn from the end point to start point in a burgers circuit is termed as the *burgers vector* $\bar{b}$.

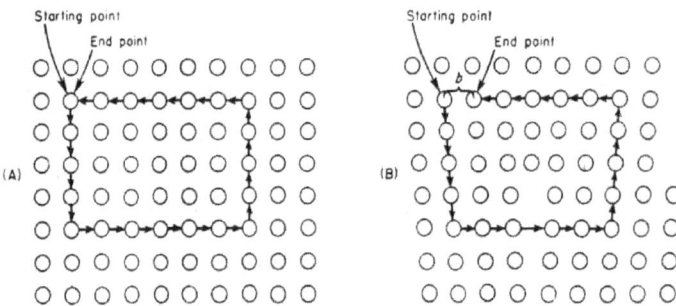

Fig.2.19. Determination of Burgers Vector.

2.31. Finding the Burgers Vector of a Dislocation:

Following rules may be observed to find burgers vector of any dislocation:

i. A circuit is traversed in the same manner as a rotating right-hand screw advancing in the positive direction of the dislocation.

ii. The circuit must close in a perfect lattice and must go completely around the dislocation in a real crystal.

iii. The vector that closes the circuit in the imperfect crystal by connecting the end point to starting point is the Burgers vector. When we make a loop, as shown in the fig.2.19(a), for a perfect crystal, we reach at the point from where we have started. However, similar loop is drawn around an edge dislocation as shown in the fig.2.20(b) we reach at a point one step ahead of the starting point. Similarly, in a screw dislocation as shown in the fig.2.20(c), if burgers circuit is drawn around the dislocation line we reach at a point one step ahead of the starting point.

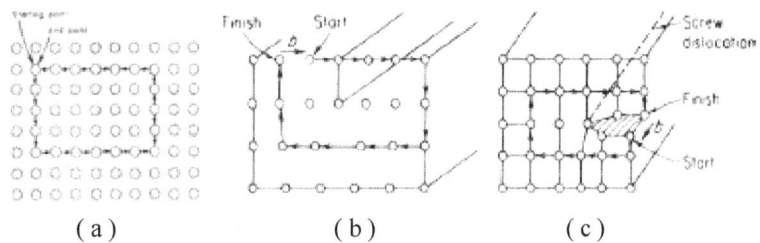

(a)   (b)   (c)

Fig.2.20. Burgers vector of : (a) Edge and (b) Screw Dislocation.

2.32. Characteristics of Dislocations:

I. Edge Dislocation:

i. An edge dislocation lies perpendiculat to its burgers vector.

ii. An edge dislocation moves on its slip plane in the direction of the burgers vector or slip direction. Under the action of a shear-stress of sense $\rightleftarrows$ a positive edge dislocation moves to the right while a negative edge dislocation moves to the left.

iii. Slip plane of an edge dislocation is uniquely defined as it contains the burgers vector and perpendicular to the dislocation line.

II. Screw Dislocation:
i. A screw dislocation lies parallel to its burgers vector.
ii. A screw dislocation moves along the slip plane in a direction perpendicular to the burgers vector or slip direction.
iii. The slip plane of a screw dislocation can be any plane parallel to the burgers vector containing the dislocation. As the burgers vector & dislocation are parallel to each other the screw dislocation can glide on any plane as long as it moves parallel to its original direction.

2.33. Force Required to move a Dislocation Line:

Let a shear stress of magnitude τ exerts a force, F on dislocation and pushes the dislocation to move through the crystal. Accordingly, force $F = \tau.A$,
where, A = Area of the slip plane and τ = Shear stress.
Peierls & Nabarro relation regarding requirement of stress to move a dislocation is: $\tau_{PN} = Ge^{-(\frac{2\pi d}{b})}$, where, G is the shear modulus, b is the magnitude of the burgers vector and d is the width of the dislocation which is the distance on either sides of the dislocation line up to which the displacements are stress-relaxing. Now it can be seen that, for narrow dislocations, $d \cong 0, \tau_{PN} = G$ and for wider dislocation, $d = 10b, \tau_{PN} = G/10^{27}$ $d = 10b, \tau_{PN} = G$.

2.34. Energy of a Dislocation Line:

As work is done on a crystal during slip the dislocations possess strain energy. The elastic strain energy may be approximated to be a hollow cylinder.

Elastic strain energy of edge dislocation: $U_{ed} = \frac{Gb^2}{2}\ln(\frac{r_o}{r_i})$

Similarly, for screw dislocation: $U_{sd} = \frac{Gb^2}{4\pi}\ln(\frac{r_o}{r_i})$, where, G is the shear modulus, r_o & r_i are the outer and inner radii respectively of an assumed elastic hollow cylinder possessing the dislocation. Further, if γ is the Poisson's ratio of the material,

we have: $\dfrac{U_{ed}}{U_{sd}} = \dfrac{1}{(1-\gamma)}$.

As a first approximation the elastic strain energy per unit length of a dislocation line with a burgers vector b is :

$$U = \dfrac{Gb^2}{2}$$

2.35. Sources of Dislocations:

The two main sources of dislocations in a crystal are:
i. Solidification of metals & alloys or the process of crystallization.
ii. Subsequent mechanical deformation or plastic deformation.

I. Frank - Read Source:

Dislocations inherently exist in crystals of metal and alloy unless special techniques are employed during their solidification from the molten state. The usual number of dislocations per cubic centimeter of a perfectly annealed metal crystal is around $10^2 - 10^3$ lines. This density increases during plastic deformation. In a highly cold-worked material, the dislocation density may be as high as $10^{12} - 10^{13}$ lines/c.c. This means the pre-existing dislocation lines in a crystal get multiplied rapidly with continued plastic deformation. The phenomenon of multiplication of dislocation lines is known as regeneration of dislocation and the dislocation line which regenerates dislocations is known as Frank-Read Source.

II. Frank - Reed Mechanism:

Frank-Reed source is a mechanism which explains the phenomenon of multiplication of dislocation lines or regeneration of dislocation line in specific well-spaced slip planes in crystals when they are deformed. When a crystal is deformed, in order for slip to occur, dislocations must be generated in the material. This implies that, during deformation, dislocations must be primarily generated in these planes. Cold working of metal increases the number of dislocations by the Frank-Reed mechanism. Higher dislocation density increases yield strength and causes work hardening of metals.

Consider a dilocation line DD' lying on the slip plne as shown schematically in the figure 2.21. The plane of the figure is the slip plane. the figure 2.21. The dislocation line leaves the the slip plane at the points D&D' as it is immobilized at these points. Immobilization may be due to formation of nodes or impurity atoms.

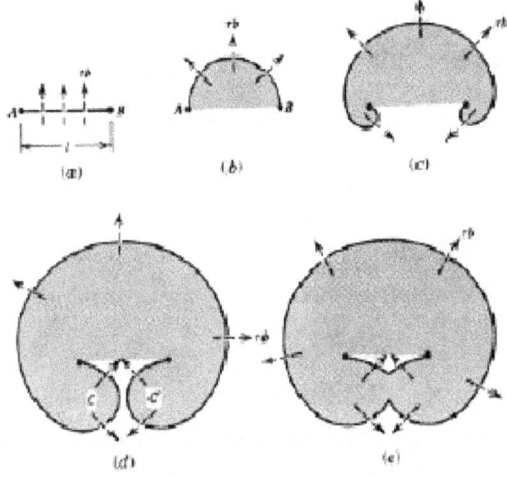

Fig.2.21. Frank-Reed Source of Dislocation Multiplication.

If a shear stress, τ acts on the slip plane, the dislocation line starts to bulge out and produces slip. For a given magnitude of shear stress the dislocation line will assume a certain radius of curveture as given by the equation $\tau = Gb/R$. The maximum shear stress required to make the bulging a semi-circle where the value of R is minimum ($R = l/2$) is $\tau_{max} = Gb/R$. Beyond this point R will increase and dislocation loop will continue to expand under a decresing shear stress as shown in the fig.2.21c. When the loop reaches a stage 2.21d. the segments m and n will meet and annihilate each other to form a large loop and a new dislocation line DD' as shown in fig.2.21e. Once the loop is detached from the pinned points DD' it expands further under increaseing shear stress and the pinned segment DD' repeats the process to produce another loop. This process is repeated over and over again producing hundreds of dislocation loops at a single source. This phenomenon of multiplying dislocations is known as Frank-Reed mechanism. However, this source does not continue indefinitely. The back stress produced by the dislocation piling-up along the slip plane opposes the applied stress and when it reaches the critical stress, $\tau = Gb/l$ the source stops operating.

2.36. Line Tension:

As strain energy is always associated with a dislocation line which is proportional to its length, work must be done to increase its length. Therefore, a dislocation line possesses a *line tension* which attempts to minimize the energy by shortening its own length. Shear stress required to convert a straight dislocation line to a curve with radius R is $\tau = Gb/2R$.

Hence, each dislocation line is associated with a force called line tension which always attempts to minimize the energy of the dislocation line by shortening its own length. For a curved dislocation line, the line tension produces a restoring force that tends to straighten it out. The line tension has the unit of energy/unit length(jm^{-1}) and is shown schematically in the fig.2.22.

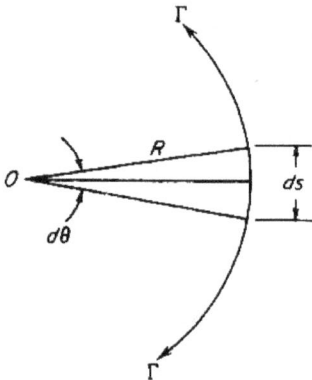

Fig.2.22. Line Tension of a dislocation line

2.37. Dislocation Pile-ups:

Dislocations frequently pile up on slip planes at barriers or obstacles such as grain boundaries, second phase particles and impurities. The dislocation pile-up produces a back stress on its source and is shown in the fig.2.23.

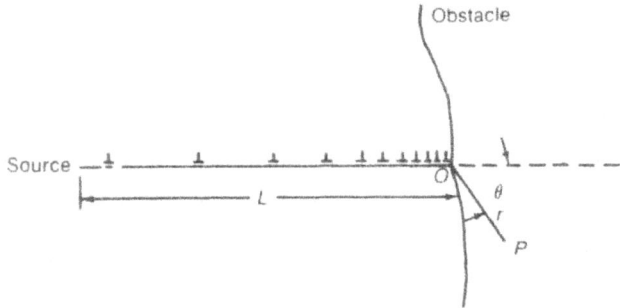

Fig.2.23. Dislocation Pile-up at an obstacle.

2.38. Deformation in Polycrystalline Materials:

In polycrystalline materials the grains are oriented randomly and the boundaries between the grains play an important role during their deformation. Single crystals deform freely on a single slip system for a sufficiently large portion of deformation and then may change its orientation by lattice rotation for continued plastic deformation. But in polycrystalline materials the individual grains are not free to deform under the applied stress as they are surrounded by other grains with completely different spatial orientation with respect to their slip systems. Further, the grain boundaries are a region of disturbed lattice and there is an abrupt change in the crystallographic orientation in passing from one grain to the next grain across the grain boundaries. In polycrystalline deformation the most important factor is the maintainance of the continuity between the grains for which the grain boundaries remain intact during deformation and voids are not created between the grains. During deformation of polycrystalline materials, each grain tries to deform homogeneously in conformity with the deformation of the specimen as a whole. The constraints imposed by continuity causes considerable differences in the deformation pattern of the neighbouring grain and also within each grain. Due to the above reason, slip may occur on non close-packed planes even at low strains in polycrystalline materials. In polycrystalline aluminium, slip occurs along $\{100\},\{110\}\,\&\,\{113\}$ family of planes instead of (111) closest packed family of planes.

Von Mises showed that for a single crystal to undergo general change in shape by slip requires at least 5-independent slip systems if constancy of volume ($\Delta V = 0$) is to be maintained during deformation. The constraints of *continuity* and *constancy of volume* leads to more complex deformation modes in polycrystalline material compared to deformation in single crystals. This led **Ashby** to suggest a dislocation model for deformation in polycrystalline materials.

2.39. Ashby's Model of Deformation in Polycrystalline Materials:

At the outset Ashby assumed two types of dislocations:

a. Statistically stored dislocations:

These are the usual dislocations available in the material as in the case of single crystals.

b. Geometrically required dislocations:

These are generated due to the non-uniform straining of grains in a polycrystalline material.

According to *Ashby's model:*

I. A polycrystalline material is deformed by disassembling its constituent grains and allowing each grain to slip independently according to Schmid's law. The process generates statistically stored dislocations leading to overlaps and voids between the grains.

II. Now each of these discrepancies, the voids and overlaps between the grains are taken in turn and corrected by introducing geometrically necessary appropriate dislocations until the grains fit together again exhibiting uniform plastic deformation on a macro scale. The order of the entire process is shown in the fig.2.23.

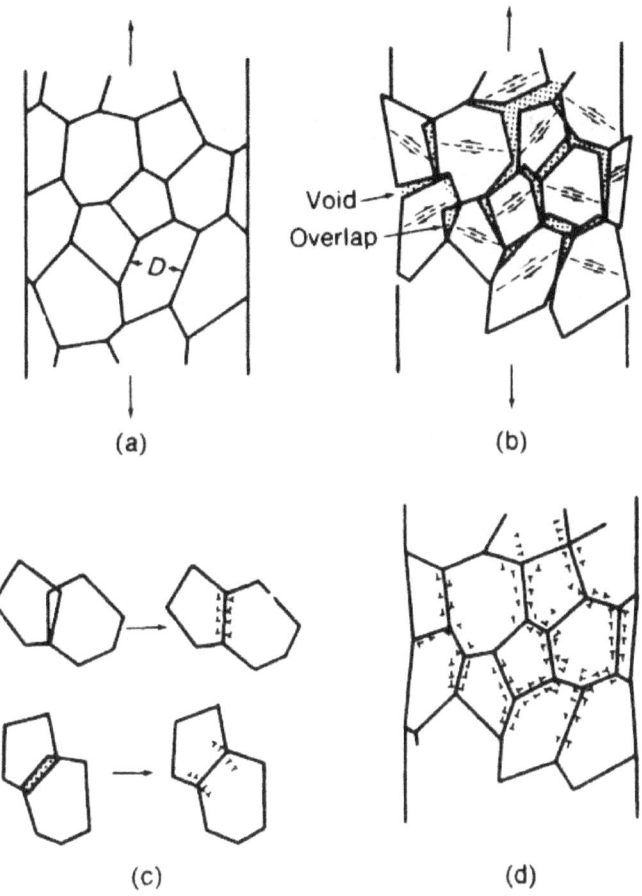

Fig.2.23. Ashby's model of Deformation of a Polycrystalline Material.

2.40. Equicohesive Temperature:

At or above a temperature of $0.5T_m$, where T_m the melting point of the metal/alloy in kelvin scale, polycrystalline materials can deform due to sliding of grains along the grain boundaries. Grain boundary sliding becomes more prominent with increasing temperature and decreasing strain rate as it happens in case of creep.

The temperature at which both grain interior and grain boundaries exhibit equal strength during deformation is defined as equicohesive temperature. Equicohesive temperature of metals and alloys is around $0.5\ T_m$, T_m being the melting point.. The following relationships hold good for polycrystalline materials at any temperature T_x:

If, $T_x < 0.5\ T_m$, grain boundary is stronger than grain interior.
If, $T_x > 0.5\ T_m$, grain interior is stronger than grain boundary.

2.41. Hall-Petch Relation:

The general relationship between *yield stress* or any other such properties and *grain size* is known as by *Hall-Petch* relation. The *Hall-Petch* relation is represented mathematically as:

$\sigma_o = \sigma_i + k/\sqrt{d}$, where σ_o = Yield stress of the material,

k = Strain hardening coefficient of the material,

σ_i = Friction stress or the overall resistance of the crystal to dislocation movement and d = grain diameter.

2.42. Yield Point Phenomenon:

During deformation, general metallic materials make a gradual transition from elastic to plastic producing a yield point on the engineering stress-strain curve. However, many alloys particularly low carbon steels show a localized heterogeneous type of elastic-plastic transition. In this case load increases steadily with elastic strain upto a particular point and then drops suddenly and fluctuates around some constant load and then rises gradually with increase in strain as shown in the fig. 2.25. As far as the stress-strain curve of low carbon steel is concerned it has an upper and lower yield point.

The elongation occuring at a constant stress is known as yielding. It is also observed that the elongation occurring around the yield point is heterogeneous. At the upper yield point a discrete band of deformed metal appears at the stress concentration points often visible to our eyes. With the formation of this band, load drops to a lower stress value known as lower yield point. Therafter the initial band of deformed metal propagates along the length of the specimen causing further plastic deformation commonly known as yield point elongation. These bands are usually at 45° to the tensile axis and are known as *Hartmann lines* or *Stretcher Strains*. This phenomenon in low carbon steels is called *Poibert effect*. Yield Point phenomenon was originally observed in low carbon steels and subsequently found in molybdenum, (α-β)brasses & aluminium. In steel, it is mainly due to the presence of carbon & nitrogen atoms.

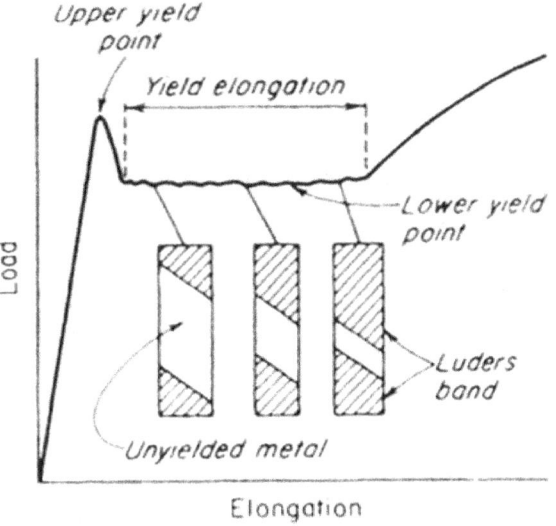

Fig.2.25. Yilding of Low carbon Steels

The onset of general yielding occurs at a stress where the average dislocation source can create slip bands through a good volume of the material. Hence, yield stress, $\sigma_o = \sigma_s + \sigma_i$, where σ_s is the stress required to operate dislocation sources and σ_i is the friction stress for movement of dislocation,

2.43. Concept of Hot & Cold Working:

Depending on temperature and strain rate of working plastic deformation of metals and alloys can be classified as:

I. Hot working
II. Cold working and
III. Worm working.

2.43.1. Hot Working:

The deformation under condition of temperature and strain rate is such that the recovery processes takes place simultaneously with the deformation is defined as *hot working*. Due to recovery large strains are achieved without strain hardening. Hot working is usually carried out above $0.6T_m$ at strain rates from 0.5 to 500s^{-1}. The lowest temperature limit of the hot working is the temperature at which the rate of recrystallization is rapid enough to completely eliminate strain hardening. Upper limit of hot working can be 50°C below the melting point of the concerned metallic material.

2.43.2. Advantages of Hot Working:

i. Hot working requires less energy for deformation.
ii. Hot working increases the ability of the metal to flow. Hence, greater extent of plastic deformation can be achieved. .

iii. Hot working helps in removing the chemical inhomogeneity in the cast ingot structures. Ingot defects like blow holes and porosities are eliminated. Coarse columnar or dendritic grains of the cast metal are broken to smaller equiaxed grains which increases the overall ductility & toughness of the cast metal.

2.43.3. Disadvantages of Hot Working:

i. Surface oxidation takes place inside and outside the furnace as the ingot is heated.
ii. There is poor surface finish of the hot worked products either due to surface decarburization or rolled in oxides.
iii. Dimensional tolerances are less in hot rolled products.
iv. Strength properties of the hot rolled products are not uniform over the entire cross-section of rolling.

2.43.4. Cold working:

When the metals and alloys are worked plastically much below their *recrystallization temperature* it is termed as *cold working*. Cold working of a metal causes increase in its strength and hardness with simultaneous decrease in the ductility & toughness. The extent of plastic deformation during cold working is for less as compared to hot working. Excessive plastic deformation will cause fracture during deformation of the stock. Hence, cold working is carried out in several steps, each step being followed by an intermediate annealing operation to soften the material before further deformation. Such an activity can achieve any degree of plastic deformation & is frequently termed as cold working-annealing cycle.

By suitably adjusting the cold working-annealing cycle, components with any degree of strain hardening can be produced. Hence, plastic deformation much below $0.6T_m$ which fails to cause any dynamic recovery or recrystallization during plastic deformation may be termed as cold working.

2.43.5. Advantages of Cold Working:

i. There is no surface oxidation or decarburization during cold rolling. Hence, the cold worked products have excellent surface finish.
ii. During cold working it is possible to simultaneously increase the strength and hardness of the rolled product to the desired level.
iii. Cold rolled products have excellent dimensional tolerances.
iv. The process is proved to be cheaper to produce smaller parts in mass scale.

2.43.6. Disadvantages of Cold Working:

1. Cold working requires more energy for deformation.
2. Cold working decreases the ability of the metal to flow.
3. Cold working does not remove the chemical inhomogeneity and cast ingot structures such as blow holes & porosities. Coarse columnar or dendritic grains are stretched further which increases the brittleness & reduces the toughness of the metallic materials.

2.44. Recovery, Recrystallization & Grain Growth:

These are solid state phenomena associated with annealing of a plastically deformed crystalline metallic materials. During cold plastic deformation, internal energy of the material increases with increasing amount of strain or deformation.

This is primarily due to increase in the dislocation density of the cold worked material over the undeformed material. Though the dislocations are mechanically stable they become thermodynamically unstable in the material. With increase in temperature, the cold worked state becomes more and more unstable which finally softens and reverts to a strain free condition. The overall process by which this occurs is known as *annealing*. Annealing is commercially important as it restores the ductility and toughness of the cold worked metal. It is possible to deform most metals to a great extent by adopting intermediate annealing operations after each step of plastic deformation. The annealing process can be divided into three distinct processes such as: *recovery, recrystalization* and *grain growth* on the basis of the microstructural changes that takes place during the entire process.

2.44.1. Recovery:

Restoration of physical properties of the cold worked metals without any observable change in their microstructure is defined as recovery. The properties related to density of the point defects in the lattice are fully recovered during this process. The temperature zone of this process is around $0.1 T_m$. The physical properties of the cold worked metal which are affected during recovery are:

:i. Internal stresses left in the material after cold working are fully relieved.

ii. Electrical conductivity of the material which has decreased after cold working is fully restored.

iii. Realignment of dislocation takes place to achieve an overall lower energy configuration.

iv. Elimination of point defects along with annihilation of some random dislocations of opposite signs occurs during recovery.
v. No perceptible change in microstructure during recovery.
vi. The strength properties which are controlled by dislocation density in the material are not affected by recovery.

2.44.2. Recrystallization (0.3-$0.5 T_m$):

Recrystallization is defined as the process of replacement of cold-worked grain structure by a set of new strain free grains of the cold worked metallic material. Recrystallization is readily detected by metallographic methods. There is no crystallographic change of the cold worked metal during recrystallization. However, all the cold deformed and distorted grains are replaced by a set of new strain free equiaxed grains. On recrystallization most of the strength properties are brought back to the values possessed by the material in the undeformed state. The property which is affected mostly by recrystallization process is the dislocation density in the material. Dislocation density decreases considerably on recrystallization and effects of strain hardening are eliminated completely. The driving force for recrystallization is the strain energy difference between the cold worked and recrystallized states of the material. In this context the Recrystallization Temperature of the material is defined as the temperature at which a given metal or alloy in a highly cold worked state completely recrystallizes in one hour.

2.45. Variables Influencing Recrystallization:

The major variables which influence recrystallization are:
i. Amount of plastic deformation prior to recrystallization.

ii. Temperature at which plastic deformation is carried out.

iii. Initial grain size of the material.

iv. Composition of the material.

v. Amount of recovery prior to recrystallization.

2.46. Laws of Recrystallization:

i. A minimum amount of deformation is required to cause recrystallization. Higher the degree of deformation lower is the recrystallization temperature.

ii. Finer the initial grain size, lower is the recrystallization temperature.

iii. Final grain size at the end of recrystallization depends primarily on degree of prior plastic deformation rather than the temperature of annealing. Greater the degree of deformation smaller are the recrystallized grains.

iv. An increase in the annealing time decreases the recrystallization temperature. Recrystallization temperature decreases exponentially with increase in annealing time.

v. Recrystallization temperature drops with the purity of the metal while alloying additions increase the same. For pure metals, $R.T. \approx 0.3 T_m$ & for alloys: $R.T. \approx 0.5 T_m$. Recrystallization slows down the in the presence of second phase particles in the alloy matrix.

2.47. Grain Growth:

Grain growth refers to the increase in the average grain size after the process of recrystallization is completed. The set of new strain free grains which are produced during recrysallization will grow on further annealing.

This phenomenon is called grain growth. The driving force for grain growth is appreciably lower than the driving force required for recrystallization. Hence, the temperature at which recrystallization occurs will also causes simultaneous grain growth. However, growth strongly depends on temperature. The driving force for grain growth comes from the reduction in the grain boundary energy per unit volume of metal. Variation in properties of metals & alloys during annealing is shown schematically in the fig. 2.26.

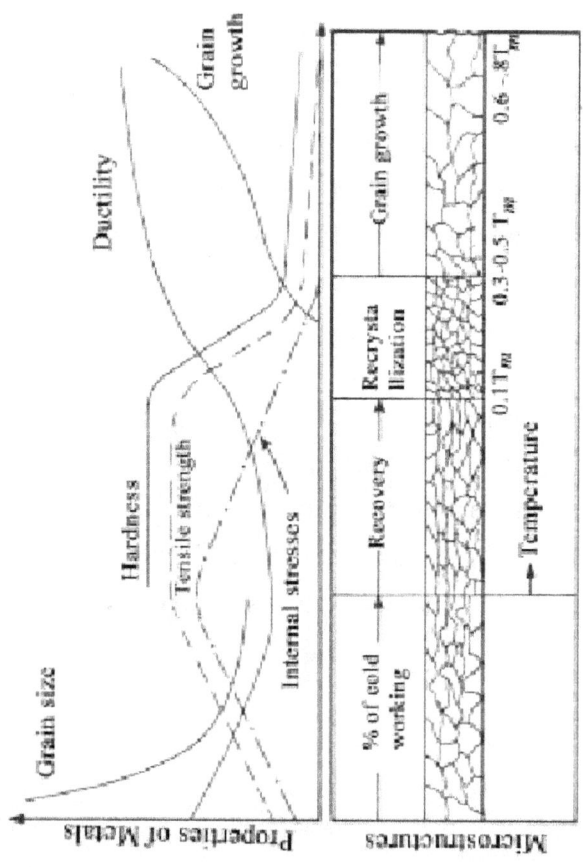

Fig.2.26. Variation in Properties and Microstructure of metals due to Cold Working, Recovery, Recrystalluization and Grain growth.

Grain growth is not a desirable phenomenon as it coarsens the recrystallized grain that leads to decreases the toughness of the material. However, grain growth can be inhibited considerably by the presence of fine dispersions of second phase particles in the matrix restricting the grain boundary movement.

2.48. Preferred Orientation:

In case of polycrystalline materials, orientation of the grains in a particular crystallographic direction(s) is sometimes desired to achieve some specific properties. Such an orientation of grains is called preferred orientation and is achieved by metal working. Metal which has undergone severe plastic deformation during rolling or wire drawing will develop a preferred orientation or texture along the direction of maximum strain as shown in the fig. 2.27.

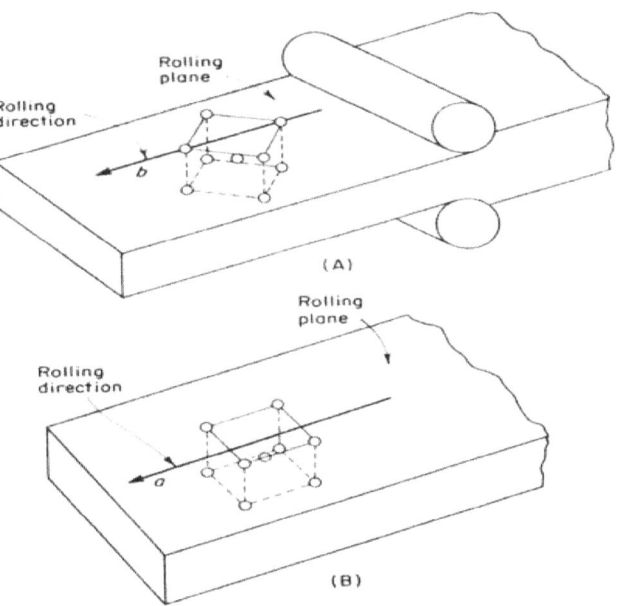

Fig. 2.27. Texturing during Rolling.

Preferred orientation can be detected by X-rays. Crystallographic fibering is brought about by crystallographic reorientation while mechanical fibering is brought about by the alignment of inclusions, cavities and second phase particles in the primary direction of the mechanical working. Both crystallographic & mechanical fibering provides directional properties to the plastically deformed materials. Such behaviour is termed as anisotropy. Anisotropism is extremely important in producing components like crane hooks and crank shaft of automobiles.

****

Chapter 3

TENSILE TEST

3.1. Introduction:

Tensile or compressive strength is the body of knowledge which deals with the relationship between internal forces, deformation and external loads. These tests are performed by Universal Testing Machine (UTM) or Tensometer. Tensometer is used for thin sections of low strength materials as it has a maximum loading capacity of 2000 kgf (@20tons) while UTMs are used for thick sections of high strength materials. UTMs can have a loading capacity as high as 980000kgf equivalent to a load of 1000tons. While conducting tensile tests few assumptions are made as precondition.

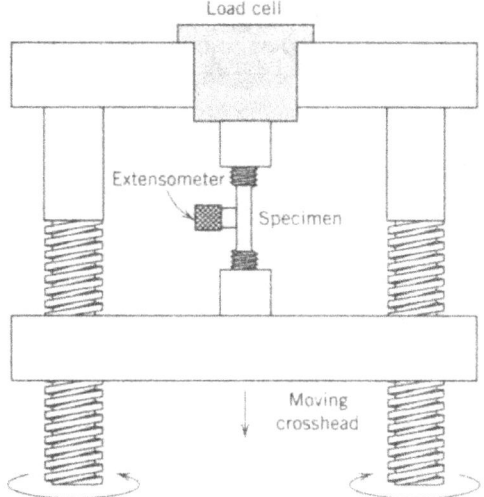

Fig. 3.1. Schematic UTM

3.2 Assumptions made during Tensile Test:

i. The block of material or the body under test is continuous. A continuous body is one which contains no voids or empty spaces within it on a macroscopic scale.

ii. The body under test is homogeneous. A body is homogeneous if it has identical properties at all points within its volume.

iii. The body under test is isotropic in nature. A body is considered to be isotropic with respect to some property when the magnitude of that property does not vary with respect to the direction & orientation of testing. If the property varies with the direction & orientation of testing the material is known as anisotropic. Most of the engineering materials satisfy the above conditions on a macro scale and considered to be continuous, homogeneous and isotropic for all practical purposes.

3.3. Significance of Tensile Test:

The tensile test evaluates several important engineering properties of the materials which include strength, ductility, elasticity, malleability & etc.

3.3.1 Strength:

A material's static strength can be defined in two ways:

i. It is the resistance of the material to permanent deformation when external force or load acts on it.

ii. It can be also be defined as the resistance to fracture. Basing on the above definitions the strength has different nomenclature under conditions of loading and extent of deformation.

3.3.2. Deformation:

Any change in the original shape & size of the material due to the action of external load is defined as deformation. Usually the metallic materials exhibit two distinct types of deformations:

i. Elastic and
ii. Plastic

I. Elastic Deformation:

It is observed that many solid materials regain their original shape and size if the load working on them is taken off within certain limit of load. This character of the materials is known as elasticity. Hence, this type of deformation is known as **Elastic Deformation**. The load upto which the material behaves elastically is known as *Elastic Limit*.

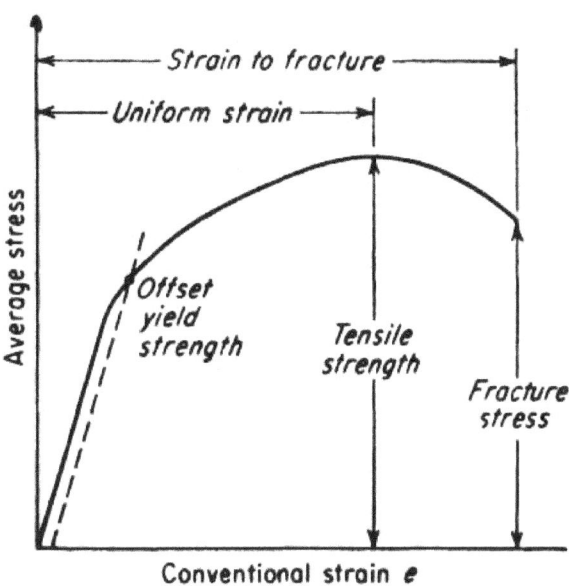

Fig.3.2. Schematic Stress-Strain diagram of a ductile Metal.

II. Plastic Deformation:

When the load working on the material exceeds the elastic limit, the material undergoes permanent deformation known as **Plastic Deformation**. Plastic deformation can be defined as a permanent deformation undergone by the material, which will not be recovered even after the removal of the deforming load.

3.4. Interrelation ship between the Load & Deformation:

The external load and the associated deformation are related to each other in all materials. For metals and alloys the deformation is directly proportional to the load. There is a linear relationship between the stress and strain within the elastic limitt. When the load is converted to stress σ & the deformation is converted to strain, mathematically we can write stress, $\sigma = E \times e$ within the elastic limit. The slope of the stress vs. strain curve, usually a straight line, is defined as modulus of elasticity(E). It is treated to be a material constant. This linear relationship between stress and strain is also known as Hooke's law. Beyond the elastic limit, strain is non-linearly related to stress.

3.5. Stress (σ):

When an external load is applied on a solid, the force acting on it tries to change the shape and size of the body. However, a force of similar magnitude develops within the body simultaneously which resists the deformation caused by the external force. This internal force which resist the deformation per unit area of cross-section is known as stress. The magnitude of stress is simply F/A, where, F is the magnitude of external force and A is the area of cross-section of the body on which this external force is acting.

Mathematically stress, $\sigma = F/A$

3.6. Units of Stress:

In SI system the stress has the unit of Newton/mm^2 while in MKS system it is kgf/mm^2. 1 kgf is equal to 9.8 Newton of force.

1kgf is the force exerted by one kilogram of mass in standard earth gravity (defined as 9.8m/s²). Often load is represented in ton which is 1000 kg of mass. Hence, a 100ton UTM machine can deliver a force of 980000 Newton or around 98000 kgf.

3.7. Strain (e):

The change in dimensions with respect to the original dimensions associated with stress is known as strain. The average longitudinal strain is expressed mathematically as $e = \Delta l / L_0$. Further, strain has no dimension.

3.8. Engineering Stress-Strain Diagram:

It is extremely important to draw a stress-strain diagram as a part of tensile test report. It is a graphical representation of the stress-strain relationship of a material loaded upto failure. The shape of the stress-strain diagram indicates the general characteristics of the material as well as its specific engineering properties. It is highly useful while designing machine components from different materials for a particular purpose.

3.8. Evaluation of Different Mechanical Properties from Engineering Stress-Strain Diagram:

A conventional stress-strain diagram for a ductile metallic material is shown in the fig.3.3. With reference to the diagram the general mechanical properties evaluated are:

I. Proportional Limit:

II. Elastic Limit:

III. Yield Point:

IV. Yield Strength:

V. Ultimate tensile strength:

VI. Elasticity:

VII. Modulus of Elasticity:

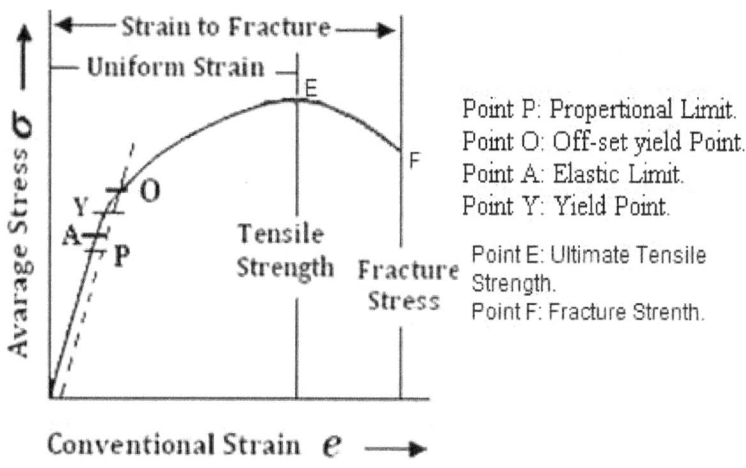

Fig.3.3. Engineering Stress-Strain Diagram.

I. Proportional Limit:

For many materials, the initial portion of the stress-strain diagram is approximately a straight line which generally follows Hooke's law. In this range any increase in stress will result in a proportionate increase in the strain. The stress at the point P is termed as proportional limit. This is the point from where the stress-strain graph starts to deviate from the linearity.

II. Elastic Limit:

Elastic limit is defined as the stress upto which the material deforms elastically and the original dimensions of the material under test are restored completely upon the release of load.

Further, there will be an onset of permanent deformation of the material if is continued to be loaded beyond this limiting stress. For most structural materials the elastic limit is nearly same as that of proportional limit. Within the elastic limit Hooke's law is obeyed. Point *A* on the stress-strain diagram indicates elastic limit.

III. Yield Point:

If the material is continued to be loaded beyond the elastic limit, a stress is reached at which the material starts deforming continuously without much increase in the load or stress. This point on the stress-strain diagram is known as yield point σ_y.

However, this phenomenon is distinctly observed only in few ductile materials like low carbon steels. In fact for low carbon steels, the stress may decrease momentarily in actual practice exhibiting upper & lower yield points. Points B and C indicate the upper and lower yield points of low carbon steels respectively. Since the yield-point is relatively easy to determine and the associated plastic deformation at the yield point is extremely small. However, this value is of great importance in designing the machine members for materials exhibiting well defined yield points.

IV. Proof Stress:

Most of the non-ferrous metal & high strength steels do not possess any well defined yield point or strength. As the yield strength is the maximum useful strength, for these materials, it is defined precisely as the stress at which the material exhibits a predefined amount of strain or deformation which can be accomadated by the engineering structure. It is graphically shown in the fig.3.4.

For all practical purposes the predefined strain for determining the yield strength is usually 0.01 or 0.02 % of the total strain upto failure of the material(points F and A respectively). The yield strength of the material determined from the engineering stress-strain diagram at the predefined strain level is termed as Off-Set Yield Point. The off-set method is explained in the figure 3.4.

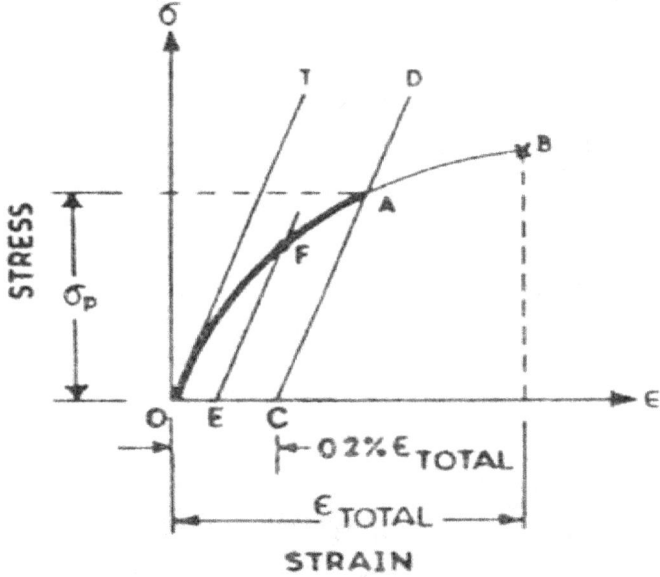

Fig.3.4.Evaluation of Proof Stress for a ductile Material

To determine the Off-Set Yield Point the amount of off-set taken is usually 2% of the total strain on the strain axis and a line is drawn up from the off-set point making it parallel to the initial portion of the stress-strain curve so as to intersect the same. From the point of intersection a perpendicular is dropped on to the stress axis so as to indicate the Off-Set Yield Point or Proof Stress.

V. Ultimate Tensile Strength:

The ultimate tensile strength of the material is defined as maximum stress borne by the specimen before it fractures basing on the original cross-sectional area. It is the maximum stress beyond which the material progresses rapidly to failure. Brittle materials fail suddenly without significant plastic deformation prior to failure while ductile materials exhibit extensive plastic deformation or necking before failure. For brittle materials the failure strength is same as that of its ultimate tensile strength while in case of ductile materials, the ultimate tensile strength is different to its failure strength. In case of ductile material the failure strength is calculated at the failure load taking the area of the necked region into consideration. In case of ductile materials, as the diameter of the specimen decreases rapidly beyond ultimate strength due to non-uniform localized necking, there is a drop-off in the load for continued deformation till the specimen fractures. The engineering stress-strain diagram shows a continuous decrease beyond the ultimate strength till the specimen fractures as the average stress is calculated over the original cross-sectional area of the specimen. The point E on the stress-strain diagram corresponds to ultimate tensile strength σ_u and point F represents fracture strength of the material(Fig.3.3).

VI. Elasticity:

It is defined as a property of the material by virtue of which the material return to its original dimension after the removal of the stress within a particular limiting stress known as elastic limit. This is useful in designing of springs or any other such components where the elastic deformation must be controlled.

Modulus of elasticity E is the measure of elasticity of the material. Higher the modulus of elasticity stiffer is the material and larger stress will be required to deform it to a particular extent. Modulus of elasticity is a structure insensitive property, slightly affected by alloying, heat treatment or cold working.

VII. Modulus of Elasticity:

Modulus of elasticity can be measured from the engineering stress-strain diagram as explained in the fig.3.5. The slope of the initial straight line portion of the engineering stress-strain diagram is defined as modulus of elasticity or Young's modulus.

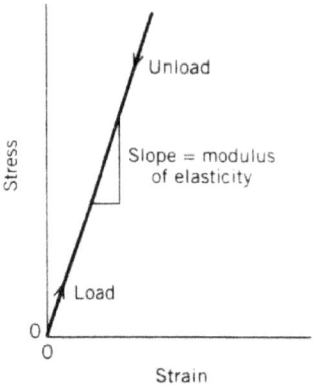

Fig.3.5. Madulus of Elasticity.

The interrelation between stress and strain is expressed by the equation: $\sigma = E \times e$ which is also known as Hooke's Law. Mathematically $E = \Delta\sigma/\Delta e$.

Modulus of elasticity is an intrinsic property of the material and is viewed as an indication of the stiffness of the material against deformation. It is expressed in kgf /mm^2.

Higher the value of E, stiffer is the material. Stiffness is an important engineering property largely considered during designing of beams & columns.

3.7. Ductility and its Measurement:

It is the property which enables the material to be formed into different shapes by rolling, forging, bending, drawing, extrusion or similar such processes. Therefore, ductility is very useful property as it indicates the extent of plastic deformation that is possible upto failure. It provides following methods to measure ductility numerically:

i. Ductility is expressed in terms of percentage elongation of the specimen over the original gauge length till the specimen fractures during the tensile test. It is calculated as follow:

The initial gauge length be, L_o and at fracture the gauge length be extended to L_f. The percentage elongation at failure is calculated to be: $\{(L_f - L)/L_o\} \times 100$ is a measure of ductility of the material. Higher the elongation higher is the ductility.

ii. Another way of expressing ductility is the percentage reduction in area at failure. For numerical evaluation let us consider the followings:

Let the initial cross-sectional area of the specimen be A_o and the area of the necked zone at failure be A_f. Now the percentage reduction in area at failure: $\{(A_0 - A_f)/A_o\} \times 100$, is a measure of ductility of the material. Higher the percentage reduction in area at failure higher is the ductility.

3.7.1. Significance of Ductility Measurement:

In general the measurement of ductility is significant in the following cases:

i. To indicate the extent upto which a metal can be deformed without fracturing during metal working operations such as rolling & forging.

ii. To indicate the plastic flow ability of the metal upto fracture.

iii. To serve as an indicator of changes in impurity level or processing condition.

3.8. Resilience & Toughness:

Resilience and Toughness are two more physical properties of the materials of great significance and importance. These two quantities can be studied with the help of the stress-strain diagram of the material as follow:

i. Resilience:

The ability of a material to absorb energy when deformed elastically and to return the same when unloaded is called resilience. This is usually expressed as **Modulus of Resilience**. It is the strain energy per unit volume required to stress the material from zero stress to elastic limit. Hence, the area under the stress-strain curve within the elastic limit represents the modulus of resilience.

Mathematically, the strain energy per unit volume of material during the uniaxial tension test can be expessed as: $U_R = \frac{1}{2}\sigma_y \varepsilon_y$, where, σ_y is the elastic limit and ε_y is the corressponding strain as shown in the fig.3.6.

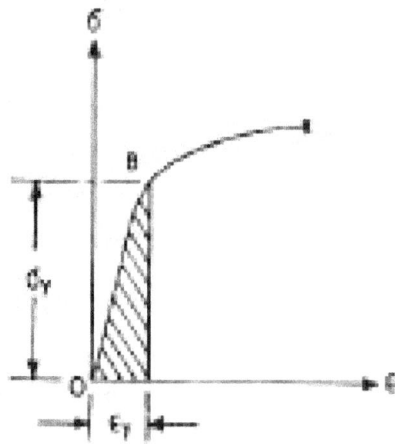

Fig.3.6. Stress-Strain diagram showing σ_y and ε_y

(ii) Toughness:

The toughness of the material is defined its ability to absorb energy during plastic deformation. The ability to withstand occasional stresses above the yield stress without fracturing is another way of defining the toughness of the material. Quantitatively toughness is the energy absorbed by the material in the plastic zone represented by the area under the stress-strain curve upto fractureas shown schematically in the fig.3.7.

Mathematically toughness, $U_T = \dfrac{\sigma_0 + \sigma_u}{2} \times \varepsilon_f$ for ductile materials and $U_T \cong \dfrac{2}{3}\sigma_f \varepsilon_f$ for brittle materials, where U_T is the fracture toughness of the material σ_0, σ_u & σ_f are the yield, ultimate & fracture strengths of the material and ε_f is the strain at fracture.

3.9. Comparison of Resilience & Toughness:

The schematic stress-strain diagram of low carbon structural & high carbon spring steels are shown in the fig.3.7.

1. On comparison it is found that spring steels have much higher resilience compared to low carbon structural steels as the area under the elastic zone for the former is higher than the later.

2. However, the total area under the stress-strain curve upto fracture is much higher for low carbon steels compared to spring steels. Hence, the toughness of low carbon steels is higher as compared to spring steels. The above facts illustrate that toughness is a parameter which comprises both strength and ductility while resilience is only related to yield or elastic strength of the material.

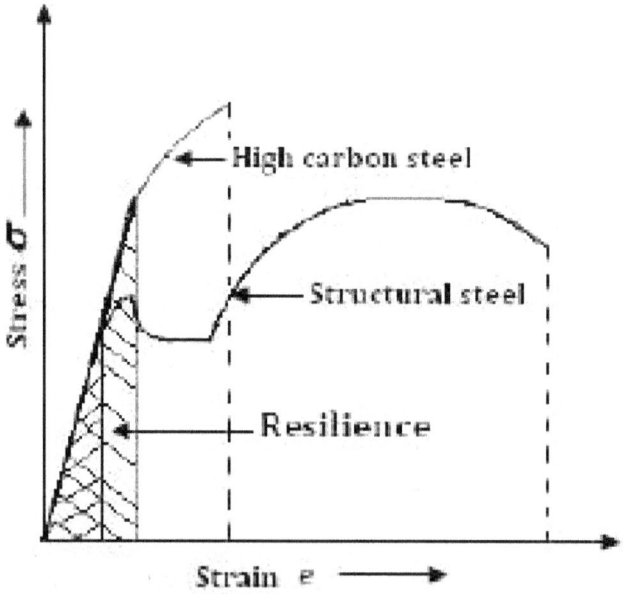

Fig.3.7. Stress-Strain diagram of low carbon structural & high carbon spring steels.

3.10. Laboratory Test Practice for Uniaxial Tensile Test:

Equipments required:

1. Universal Tensile test Machine (UTM).
2. Micrometer (0-25mm dial gauge).
3. Vernier caliper.
4. Extensometer.
5. Specimen Griping Device.
6. Specimen

Specimen:

I. Circular Section:

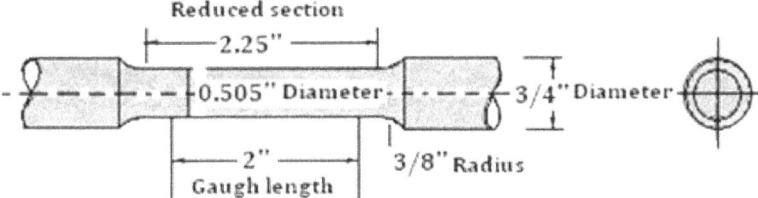

II. Flat Section:

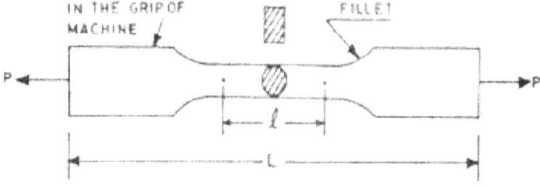

Set up:

While conducting this test, the most important parameter is the gauge length, which depends on the geometry of the test specimen. Gauge length L_o for differnt specimen is taken as per the following relationship with crosssctional area A_o:

i. $L_o = 5.65\sqrt{A_o}$, for circular sections and

ii. $L_o = 4\sqrt{A_o}$, for rectangular strips.

3.11. Standard ASTM & BIS Tensile Specimen:
I..Sheets and Flats:
$L_o/A_o = 4.5$.

Standard specimen:

$L_o = 50mm$ and $A_o = 2 \times 2mm$

II. Round Bars, Rods and similar shapes:
$L_o/D_o = 4.0$

Standard specimen:

$L_o = 50mm$ and $D_o = 12.8mm$

III. Specimen Grips:

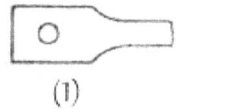
(1)
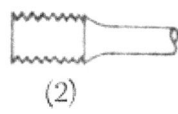
(2)
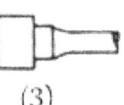
(3)

3.12. Test Method:

After marking proper gauge length, the specimen is gripped firmly between the jaws of the UTM vertically. The machine is adjusted to zero load. Then the extensometer is attached to the sample with zero reading. Then the tensile load P is applied gradually in steps of 250kg. For each increment of load, the increase in the gauge length Δl is measured precisely by extensometer. This incremental loading in steps of 250kg is continued till the yield point is reached and is recorded. At this point the extensometer is removed and loading is continued. The change in length beyond yield point is measured by a precision scale till the specimen fails with a thud sound. The final gauge length, surface area of the fracture are determined by placing both pieces in proper contact.

Load versus elongations recorded are plotted to generate the stress-strain diagram. The stress-strain diagram can be utilized for finding out the followings:
1. Tensile yield point or yield strength.
2. The ultimate tensile strength.
3. Proof stress.
4. Fracture strength.
5. Ductility.
6. Elastic modulus.
7. Nature of the material like ductile, brittle etc.

3.12. Comparison of Different Materials with respect to their True Stress-Strain Diagrams:

According to the stress-strain diagrams materials in general can be classified as (refer fig.3.8.):
I. Ductile: Copper, Aluminium, Gold & etc.
II. Brittle: Cast Irons, Glasses, Ceramics & etc.
III. Ductile-Brittle: HY Steel, High carbon Steels & etc.
IV. Rubber & Plastics: Superplastic materials.

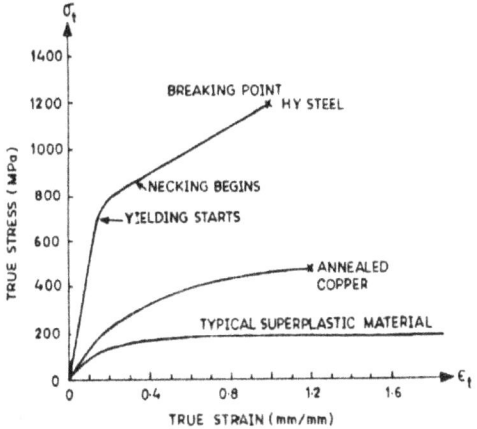

Fig.3.8. True Stress-Strain diagrams for Different Materials.

3.13. Compression Test:

In this case the specimen can be either cubes or cylinders shown in the fig.3.78. As the compression test suffers from end restrain and lateral instability, the specimen may bulge out laterally as shown in the figure. Further it may buckle when the specimen is too long with respect to the area over which it rests during the test. The most suitable value of h/d ratio is taken to be ≤ 2. This means the shape and size of the specimen has an effect on the compressive strength which decreases with an increase in h/d ratio. This test is carried out on brittle materials like cast irons, concrete blocks.

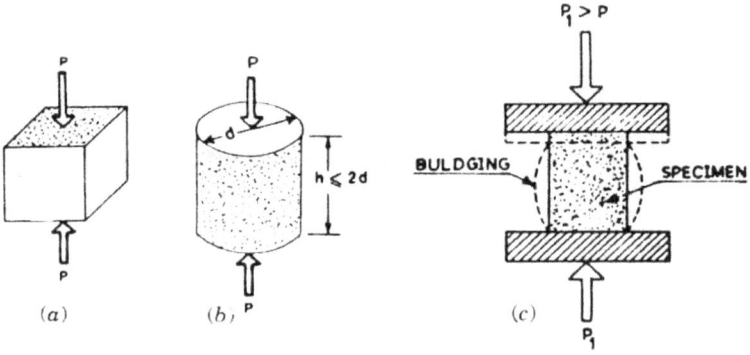

Fig.3.9. Compression Test Specimen: (a) Cube; (b) Cylinder; (c) Bulging Action during Compression Test.

3.14. Comparison of Compressive σ_c & Tensile σ_t strengths:

Materials	Compressive Strength (σ_c) kg/mm²	$[\sigma_c/\sigma_t]$
Grey cast	800	5.0
Concrete	35	12.5
Alumina	2000	10.0
Plexiglass	115	1.5

The table indicates higher comprssive strengths for brittle materials compared to their tensile strengths.

3.17. A Note on True Stress & True Strain:

The engineering stress-strain diagram does not give a real indication of the deformation characteristics of the materials as it is entirely based on the original dimensions of the specimen which change continuously during the test. Hence, it is important to measure the stress and strain based on the instantaneous dimensions of the specimen during the test. The instantaneous values of stress & strain based on the changed dimensions are known as true stress and strain. A schematic comparision between true stress-strain and engineering stress-strain in a material is shown in the fig.3.10.

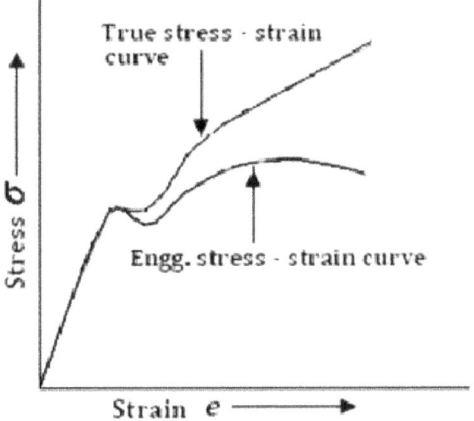

Fig.3.10. True and Engg. Stress-Strain Diagram of mild steel.

i. True Stress:

True stress is the load at any instance to the instantaneous cross-sectional area over which it works. The conventional engineering stress is calculated over the original cross sectional area. There is almost no difference between the true and conventional engineering stress in the elastic zone but they start to differ in the plastic zone just after the yield point.

ii. True Strain:

As per definition true strain, ε is defined as the change in length to the instantaneous gauge length. It is the change in linear dimensions divided by the instantaneous value of the dimension. The concept of true stress and strain was first proposed by Ludwig.

ellustration:

A specimen of gauge length 5cm. is elongated to 6cm. in steps of 0.1cm in each second. Find out the engg.& true strain.

Solution: Engineering Strain, $e = \dfrac{\Delta l}{L_o} = \dfrac{6-5}{5} = \dfrac{1}{5} = 0.20$

True strain, $\varepsilon = \ln[\dfrac{L_f}{L_o}] = \ln[\dfrac{6}{5}] = 0.182$

3.17. Relation between True Strain ε and & Engg. Strain e :

Engineering strain, $e = \dfrac{\Delta l}{L_o} = \dfrac{L_i - L_o}{L_o} = \dfrac{L_i}{L_o} - 1$.

or, $(e+1) = [L_i/L_o]$ -------------(1)

Taking ln of both sides, we have, $\ln(e+1) = \ln[L_i/L_o]$ --------(2)

As per definition $\ln[L_i/L_o] = \varepsilon$.

Hence, $\ln(e+1) = \varepsilon$ ----------(3)

3.18. Relation Between Engg. Stress S and True Stress σ :

Engineering stress, $S = P/A_o$, where, A_o is the initial cross-sectional area and P is the load working.

Hence, True Stress, $\sigma = P/A_i$

where A_i is the instantaneous cross-sectional area.

Applying principle of constancy of volume relationship we have: $A_0 L_0 = A_1 L_1 = ---- = A_i L_i = ---- = A_f L_f$

or, $\dfrac{A_0}{A_f} = \dfrac{L_f}{L_0}$

From equation (1) we have, $(e+1) = \dfrac{L_f}{L_0}$.

Now, $\sigma = \dfrac{P}{A_f} = \dfrac{P}{A_0} \times \dfrac{A_0}{A_f} = S(e+1)$ -------------(2)

****

Chapter 4

HARDNESS TEST

4.0. Introduction:

It is quite difficult to define the hardness of a material as it is not related to any intrinsic property of the concrned material. Hardness is usually expressed as a number depending upon the particular test method used to determine its value. Hardness values cannot be used directly in design activities like tensile strength. However, it is important to evaluate hardness as it is related to both elastic & plastic properties of the material. The test values can be used to estimate other mechanical prosperities; most significantly the tensile strength. Hardness test is widely used for routine inspection & quality control of steel components subjected to heat treatments as the same is widely affected during heat treatments. When hardness resulting from a set of particular heat-treatment practices is well established, the hardness test affords a very rapid and simple means of inspection and quality control of the engineering materials.

4.1. Classification of Hardness:

Hardness usually implies resistance of the material to plastic deformation. For metals and alloys this is a measure of their resistance to plastic or permanent deformation. Further, hardness of a material can also imply other properties primarily related to its surface condition. For simplicity hardness of a material is classified into different categories depending on the test method adopted.

I. Elastic Hardness:

The resistance offered by the metal to impact and the resultant rebound of the impact is termed as elastic hardness. This hardness is a relative measurement of elasticity of the material. The device used to measure this harness is called **Scleroscope**. The arrangement in this device is such that it measures the rebound height of a small diamond tipped hammer that is allowed to drop from a definite height onto the metal surface by its own weight, after hitting the surface it rebounds. The rebound height is taken as a measure of the surface hardness. Higher the height of rebound higher is the hardness. Such hardness measurement is also called **Rebound** or **Shore Hardness.**

4.2.1. Principle of Scleroscope:

The Scleroscope has a linear hardness scale against which the rebound height of the hammer is measured. The total scale length is equal to the rebound height for a quench hardened high carbon steel surface taken as standard. This rebound height is divided into 100 equal divisions. For wider application the scale is extended upto 104 divisions to measure relative hardness of the very hard brittle materials & mineralogical crystals. The basic principle of measuring relative hardness is the rebound height form surface of the material under consideration.

The Scleroscope contains the hammer with a metal plug that holds the diamond indenter. The indenter weight only 2.6 grams. The hammer strikes the test piece at the requisite area and momentarily exerts a stress of around 470 tons per sq.inch which is sufficient to produce a permanent indentation.

Marring effect of indentation even on the best finished surfaces may be disregarded for its negligible size. Kinetic energy associated with the falling hammer at the instant when contact is established with the specimen surface is mainly expended in forming the indentation in the elastic range and the remaining energy is utilized in rebounding the hammer. In general, softer the material larger will be the indentation. As more energy is expended during the test in forming the indentation it results in shorter rebound height. Reversely, for harder materials less energy is expended in forming the indentation resulting in higher rebound height of the hammer.

4.2.2. Application of Shore Hardness Test:
1. This is mainly used as a portable hardeners tester to test the harness of the material in situ. It has been most useful in testing hardness pipe lines and machine beds.
2. Scleroscope can be used to measure the hardness of harder materials which are as thin as razor blades(0.15-0.16mm) and in the range of 0.25-0.30mm for softer material like brass & copper.
3. Hardness of the irregular surfaces can also be estimated without error proper clamping device to make the rebound perfectly vertical.

4.2.3. Precautions to be taken during Rebound Harness Test:
Most of the technical errors during rebound test are taken care of by using proper clamping devices and perfect surface preparations as to allow the hammer to hit the surface at right angle.
1. The surface of the material must be prepared and made perfectly flat without any oxidation or decarburization prior to testing.
2. Proper clamping devices should be used to avoid vibration.

3. Hardness checking at a single point should not be repeated as it will show errors or may cause chipping (breaking) of the diamond tip of the hammer indentor.

4. Multiple tests are to be carried out at different locations of the surface for multiphase materials containing two or more phase with large differences in their hardness values to obtain fairly accurate average hardness value.

4.3. Scratch Hardness Test:

One of the systematic hardness scales ever devised was first proposed by the scientist Friedrich Mohs. This scale provides a means by which the hardness of the material can be determined quickly and effectively by scratching the material with a series of standard materials. Though this method is rarely used by the metallurgist, it finds extensive application in the field of mineralogy where determination of relative hardness of rocks, minerals & ores is highly essential for their characterization.

4.3.1. Principle of Scratch Hardness Test:

Soft materials like talc or soap stone can be scratched by our finger nail. This means finger nail is harder than such a stone. Similarly quench hardened steel can scratch soft steel, copper, aluminium and many more materials. Hence, during the scratch test, resistance of a material to wear, abrasion or scratch is measured relatively. In this system of measurement, the scale consists of ten (10) standard minerals arranged in the order of their increasing hardness known as Mohs scale. Each mineral is being numbered according to its position in the series.

In the standard Mohs scale, Talc is at serial no.1 while diamond occupies the 10th position indicating talc being the softest and diamond being the hardest possible material. Diamond is considered to be the hardest substance available on earth as it can scratch all other materials. In fact, diamond tipped pen is used to cut glasses.

4.3.2. The Complete Mohs Scale:

Sl. No	Mineral	Relative Scratch Hardness Number
1.	Talc	1
2.	Gypsum	2
3.	Calcite	3
4.	Fluorite	4
5.	Apatite	5
6.	Orthoclase	6
7.	Quartz	7
8.	Topaz	8
9.	Corundum	9
10.	Diamond	10

4.3.3. Calculation of Hardness using Mohs Scale:

Suppose a mineral gets scratched by Apatite at no.5 but able scratches Fluorite at no.4. Then the scratch hardness value that can assigned to the mineral is in between 4 & 5.

4.3.4. Advantages of Scratch Hardness Test:

1. This test is simple, low cost and requires no technical skill.
2. This tests is fairly rapid & quick.
3. The hardness scale is only form 1 - 10.

4.3.5. Disadvantages of Scratch Hardness Test:

i. This test is finds very little use with metallurgists.

ii. The hardness scale is highly limited to find the exact hardness.

iii. The primary lacuna of this test is that, the scale used for determination of hardness is non uniform.

4.4. Indentation Hardness:

This hardness test is used widely because of its relation to the strength of the materials. In this case hardness of a material is defined as its resistance to indentation (plastic deformation) or penetration of the indenter of known geometry with a predefined static load by holding the specimen rigidly on a platform.

Depending on the type of test performed, hardness of the material is expressed by means of a number that is either:

i. Inversely proportional to the depth penetration of the indenter.

ii. Proportional to the mean load which acts over the area of indentation.

For softer materials, higher will be the depth or area of indentation reflecting a lower hardness value for the material. Reversely for harder materials, both depth and area of indentation will be lower reflecting a higher hardness for the material. The hardness value expressed in number find extensive use with the metallurgists and design engineers, as it relates to the tensile strength of the material in some classical way. The major tests under the indentation category are:

I. Brinell Hardness Test.

II. Rockwell Hardness Test.

III. Vicker's Knoop Hardness Test.

4.5. Brinell Hardness Test:

The first most widely accepted and standardized indentation hardness test was proposed by J.A Brinell in the year 1900. This test utilizes a 10mm hardened high carbon steel ball indenter which is forced into the surface of the test piece with a static load of 3000kg. An indentation is produced on the specimen surface as a result of the applied load. Mean diameter of indentations is of prime importance in measuring the hardness number. The indenter may change depending on the materials being tested. A basic sketch of brinell test is shown in the fig.4.1.

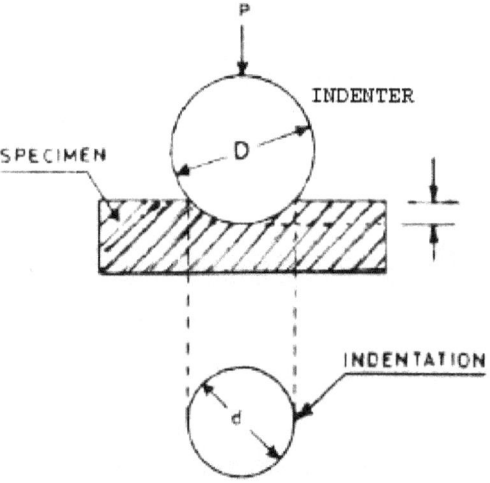

Fig.4.1. Basic Sketch of Brinell Test

4.5.1. Types of Brinell Indenters:

i. Hardened steel ball of 1, 2.5, 5 and 10 mm. size.

ii. Tungsten Carbide (WC) ball is used for very hard materials to minimize the distortion of the indenter(upto 440-625HB).

4.5.2. Principle of Brinell Hardness Test:

The load, P is applied on the surface of the test piece by the lever arrangement. For harder ferrous material like steels a load of 3000kg is selected and for softer materials like copper, aluminium alloys a load of 500 kg is usually selected to avoid deep impressions. The standardized load is applied for a specific time period to produce a perfect impression. For ferrous metals the load is applied for 10-15 seconds while for softer nonferrous materials the load is applied for a period of 30 seconds. If the test load is removed before the scheduled time period smaller indentations are produced and it gives erroneous results regarding hardness of the material. The above condition reflects a higher hardness value than the actual hardness and vice versa. The ball indenter produces an circular indentation as shown in the fig.4.2.

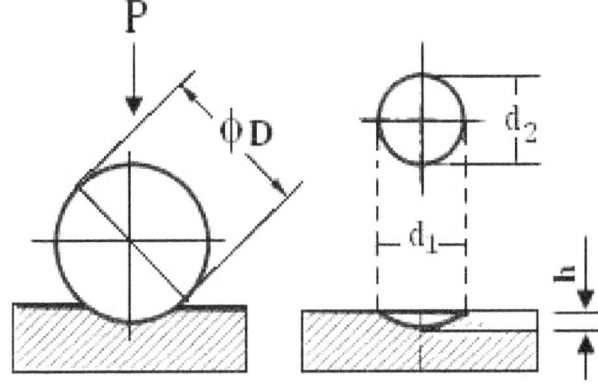

Fig.4.2. Shape and Depth of Indentation in a Brinell Hardness Test.

The diameter of the impression is measured with a low power microscope(25X) containing an ocular scale. The ocular least count is 0.05mm. The indenter impression is measured in two perpendicular directions to calculate the average diameter.

The Brinell hardness number BHN is expressed as the load, P divided by the curved surface area of indentation.

Mathematically, BHN = $\dfrac{2P}{\pi D[D-\sqrt{D^2-d^2}]} = \dfrac{P}{\pi Dh}$, where

P = Test load in kg.
D = Diameter of the indenter in mm (1, 2.5, 5 or 10).
d = Diameter of the indentation or impression in mm.
h = Depth of the indentation in mm.
The unit of BHN is kgf/square millimeter (1 kgf/mm^2 = 9.8 Mpa).
Hence BHN is a pressure term.

4.5.3. Geometry of Indentation:

As per definition, BHN = $\dfrac{\text{Applied Load (P)}}{\text{Curved Surface area of Indentation}}$

Surface area of the segment of the sphere is πdh where, h is the depth of indentation, d is the diameter of indentation & D is the diameter of indenter.

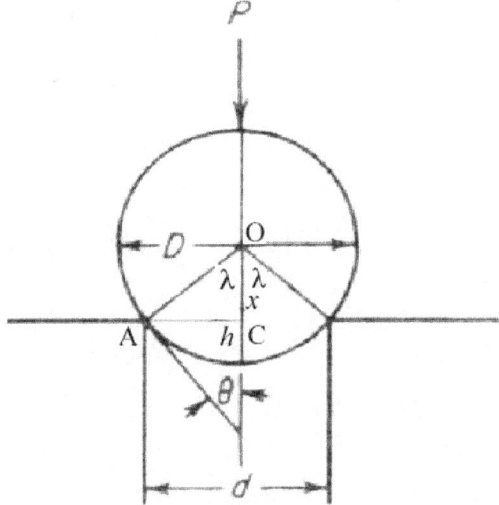

Fig.4.3. Geometry of Brinell Indentation.

From the indentation geometry (fig.4.3) we have:

$x + h = D/2 \;-----(1)$

Considering the right angled triangle OAC, $(D/2)^2 = x^2 + (d/2)^2$

or, $D^2/4 = x^2 + d^2/4$ or, $x^2 = D^2/4 - d^2/4$

or, $x = \frac{1}{2}\sqrt{D^2 - d^2} \;----(2)$.

Hence, $h = \frac{D}{2} - x \;----(3)$

So, $\text{BHN} = \dfrac{2P}{\pi D[D - \sqrt{D^2 - d^2}]} = \dfrac{2P}{\pi D[D - \sqrt{4x^2}]} = \dfrac{2P}{\pi D[D - 2x]}$

or, $\text{BHN} = \dfrac{2P}{\pi D[2h]} = \dfrac{P}{\pi D h}$. Similarly, considering the triangle OAC:

$\sin \lambda = \dfrac{D/2}{d/2}$, or, $D = d \sin \lambda$. Substituting this value of D in the eq(3)

we have: $\text{BHN} = 2P/\pi D^2 (1 - \cos \lambda) \;-----(4)$

The depth of indentation is: $h = F/(\pi D \times \text{BHN}) \;----(5)$

Where, F is the force applied in Newton (1kgf = 9.807 N),

$\text{BHN} = 2P/\pi D^2 (1$

4.5.4. Load – Indenter Diameter Interrelation:

In order to obtain similar BHN values with non-standard load or indenter, it is necessary to produce geometrically similar indentation, which is obtained by using standard load and indenter combination. Geometrical similitude is achieved so long as the included angle 2λ (as shown in the fig. 4.3) remains constant. For the included angle 2λ and BHN to remain constant, the load. P and indenter diameter, D must be varied in the ratio of:

$$\frac{P_1}{D_1^2} = \frac{P_2}{D_1^2} = \frac{P_3}{D_3^2} = \;-----(1)$$

EXAMPLE:

If the indenter diameter is changed from 10mm to 2mm what will be the revised load assuming that for 10mm indenter the load applied is 3000kg.

Solution: Using the relation: $\dfrac{P_1}{D_1^2} = \dfrac{P_2}{D_1^2} = \dfrac{P_3}{D_3^2} = -----(1)$

we have: $\dfrac{3000}{10^2} = \dfrac{P_2}{2^2}$

So, $P_2 = \dfrac{3000 \times 4}{100} = 120 Kg.$

Hence, the revised load with 2mm dia indenter is 120kg.

Note: Use $\dfrac{P}{D^2} = 30$ for steels, 10 for nonferrous alloys and 2 for very soft materials like lead.

4.5.5. Method of Expressing BHN:

The standardized conditions of BHN test are:

i. Load - 3000kg (or 3000kgf).

ii. Indenter diameter - 10mm.

iii. Time duration of the applied load - 10 Seconds.

Under the above condition the BHN is simply represented by a number. However, BHN for using any other combinations of load, indenter diameter & time other than standard condition should be indicated by an index to indicate the test conditions in the order of indenter diameter/load/duration of loading.

For example, 250BHN 2/120/15/ indicates a Brinell number of 250 using 2mm diameter indenter under the test load of 120kg applied for 15 seconds. Sometimes BHN is also written as HB.

4.5.6. Requirements of Brinell Hardness Test:

I. Indenter:

The diameter of the indenter should be as large as possible in order to test the largest representative area of the test piece. In Brinell hardness test the indenter used should be such that the d/D ratio should lie between 0.25 - 0.50 otherwise different indenter should be tried.

II. Surface Condition of the Test Specimen:

The top & bottom surfaces of the test specimen should be perfectly flat, smooth & parallel. The test surface should be free from oxides, dirt, oil and foreign materials.

III. Surface preparation of the Test Specimen:

During the surface preparation it should be ensured that the surface hardness is not changed. Rough surface should be ground flat before testing.

IV. Thickness of the sample:

The thickness of the test specimen should be at least ten times the depth of indentation.

The depth of indentation is: $h = F/(\pi D \times BHN)$.

Where, F is the force applied in Newton (1 kgf = 9.807 N), BHN is the Brinell hardness number, D is the diameter of the indenter & d is the diameter of indentation. After the test, no bulge should be visible on the opposite surface of the specimen.

V. Spacing between the Indentations:

i. The distance between the edge & centre of any indentation should be at least 3-5 times the mean diameter of the indentation.

ii. The distance between the centres of any two adjacent indentations should be at least three (3) times the mean diameter of the indentation. smooth, free from dirt & oil.

VI. Anvil:

The anvil surface in contact with the test piece should be smooth, free from dirt & oil.

VII. Test Temperature:

The test should be carried out within 10-35°C.

4.5.7. Test Procedure:

1. Selection of appropriate test force & indenter diameter.

2. Placing of the test piece on suitable anvil so that the test piece does not move at all during testing.

3. The indenter is brought into contact with the test surface by moving the anvil head.

4. The test load is then applied at a particular point perpendicular to the test surface for 10-15 second unless otherwise specified.

It should be ensured that there is no vibration or shock during loading. After the loading time is over, load is removed and the diameter of the indentation is measured in to direction right angle to each other. The mean diameter is found out. By the given formula Brinell hardness number is calculated.

4.5.8. Precautions taken During Brinell Hardness Test:

1. Surface of the test specimen should be smooth and free from oil, dirt, oxides & other surface defects.

2. The minimum thickness of the specimen should be at least 10 *times* the depth of indentation.

3. The distance between edge of the specimen and centre of any indentation should be at least 3-5 times the diameter of the impression.

4. The distance the centre of two adjacent impressions should be at least 2.5 times the diameter of the impression.

5. Round surfaces should be ground flat before testing. V–blocks may be used to hold the specimen properly. Impression should be made on the flat surface.

6. Flattening (deformation) of the spherical steel indenter takes place particularly while testing materials of BHN value 400 or more. In case of high hard materials Tungsten Carbide indenters may be used.

7. Static load is important. Impact loading is highly objectionable as it may damage the indenter.

8. If the diameter of indentation is found to be very close to 10 mm, the calculated BHN is erroneous. In such cases lower load may be used for determining hardness.

9. The load should be allowed to act only for a specified time period. It should not be more or less. If the time duration is high, the estimated hardness will be low. Reversely, if the time duration is less, the estimated hardness value will be abnormally high.

10. Further drawbacks of Brinell hardness test are:

i. Sinking effect occurs in manganese and austenitic steels. This is primarily due to higher hardness of the material being tested compared to the indenter hardness.

ii. Piling-up effect which occurs in lead, tin and magnesium alloys. This is primarily due to much lower hardness of the material being tested compared to the indenter hardness. Both the effects are shown in the fig.4.4. Hence, Brinell hardness test should not be carried out on such materials.

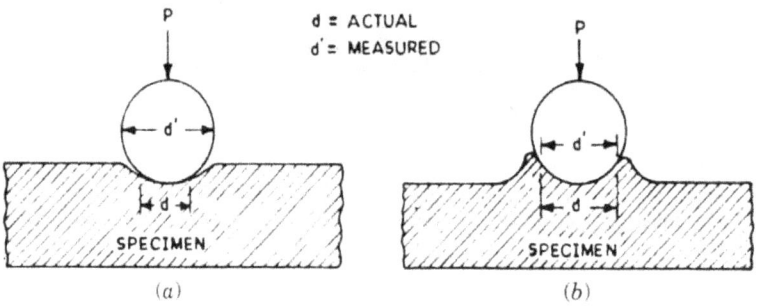

Fig.4.4. Sinking-in and Pilling-up Effects

4.5.9. Hardness Range and Load stages in BHN Test:

Materials	Hardness Range BHN (kgf/mm²)	Load Stages, P in kgf
Soft iron and Steel castings.	67- 500	30D²
Light alloys and Brasses & bronzes.	22 -315	5D²
Al, Mg & Zn alloys and cast brasses.	11 -158	5D²
Bearing metals.	6 - 78	2.5D²
Lead, Tin & Soft solder.	3 - 39	1.25D²
Soft metals at higher temperature	1 - 15	0.5D²

4.5.10. Merits & Demerit of Brinell Hardness Test:

I. Merits:

1. Due to large size of the impressions in a Brinell hardness test, the effect of local inhomogeneity is averaged out during the test.
2. This test is less influenced by the surface conditions and extensive surface preparation is not required as in other hardness test.

II. Demerits:

1. It is not possible to cover the entire range of commercial materials with a single load indenter diameter combination.
2. As the Brinell impression is large, the impression works as a potential site for failure for small and critically stressed machine component.
3. Higher technical skill is required to evaluate BHN value
4. This test is slower compared to other hardness tests.

4.6. Rockwell Hardness Test:

This is the most widely used hardness test in United States and has been accepted worldwide due to the following reasons:

i. Testing speed is much higher compared to other test methods.
ii. This test is free from personal errors.
iii. This test has the ability to distinguish very small hardness differences in hardened steel.
iv. Due to smaller size of indentation, the finished components can be tested as a quality check without damaging the same.
v. The test use different Scales for different materials enabling easier hardness comparision.

4.6.1. Principles of Rockwell Hardness Test:

It is an indentation hardness test. The test uses either a diamond cone having an included angle of 120° and 0.2 mm radius of curvature at the tip or hardened steel ball of 1.6 or 3.2 mm dia as indenter. The indenter is forced into the surface of the test piece in 2 steps by applying a minor and a major load successively.

The permanent depth of penetration of the indenter under major load is measured from which the Rockwell hardness number is derived. Rockwell hardness is expressed as:
$N = \frac{h}{S}$, where, h is the permanent depth of penetration in mm after removal of major load and S is the scale unit, which is 0.002mm for A, B, C, D &E scales.

4.6.2. Rockwell Hardness Test Method:

The test is carried out by a slowly raising specimen on an anvil against the indenter until a minor load of 10kg is applied initially which is indicated by the dial. After stabilization of the minor load the major load is applied for around 6 seconds so as to make a deep indentation. The test utilizes the depth of indentation under constant load as a measure of hardness. The depth is automatically indicated on a dial gauge in terms of an arbitrary hardness number after removal of major load. The dial contains 100 divisions where each division represents a penetration of 0.002mm. The dial recorder is reverse one. For softer materials, where the depth of indentation is large it will record a lower hardness value while for a harder material where penetration is low it will record a higher hardness.

4.6.2. Designation of Rockwell Hardness:

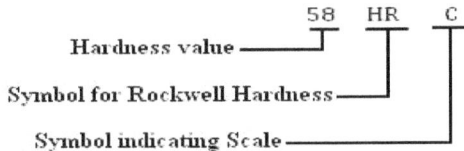

Unlike Brinell or Vickers hardness numbers having a unit of kgf/mm² Rockwell hardness number is dimensionless.

4.6.4. Types of Indenters:

1. The general indenters are 1.6 & 3.2 mm hardened steel balls.
2. A 120° diamond cone with slightly rounded tip called Brale indenter is used for hard materials.
3. The major load of 60, 100 & 150 kgs are normally used on the tester with a minor load of 10 kgs.

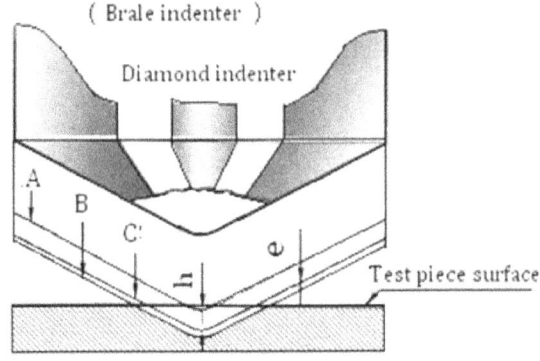

Key:

A = Position of indenter under minor load.
B = Position of indenter under total (minor + major) load.
C = Position of indenter minor load after removal of major load.
h = Permanent depth of indentation under minor load.
e = Elastic recovery just after removal of major load.

Fig.4.5. Brale Indenter

4.6.5. Scales used in the Rockwell Hardness Testers:

It is important to note that a single combination of load and indenter will not produce satisfactory result for wide range materials of different hardness. Hence, different indenters & major loads are used for determining hardness different materials. A particular combination of indenter & major load signifies a particular scale. Accordingly there are three different scales as discussed belolow:

A Scale:

This scale uses Brale indenter with 60kg major load. This scale provides the most extended Rockwell hardness scale. This scale can be used for extremely soft materials like annealed brass to extremely hard materials like cemented carbides (WC, TiC etc) materials. The hardness value is represented as $R_A 50$, $R_A 60$.

B Scale:

This scale utilizes hardened steel ball indenter with 100kg major load. The scale range is from $R_B 0$ to $R_B 100$. This scale is widely used for non-ferrous, low & medium carbon steels and other materials which are usually softer.

C Scale:

This scale was Brale indenter with 150kg major load. The hardeners values are usually expressed as $R_C 20, 40, 70$ & etc. The most useful range of the scale is from 20 to 70. This scale is most useful for hardened steel components.

4.6.6. Reasons of using Minor Load during the Test:

The total load applied during Rockwell hardness test is same as that of the major load. However, the total load is applied on the test specimen in two stages in the form of:

1. Minor load and
2. Major load.

During Rockwell test a minor load of 10kg is applied on the test surface. Once equilibrium is established, the dial guage is set to zero and then remaining load is applied on the test specimen as major load depending on the scale chosen. In case of B scale the minor load is 10kg while the major load is 90kg making the total load as 100kg. Application of 10kg minor load is common to all scales of Rockwell hardness testing excepting Superficial hardness testing where the minor load is only 3kg.

Application of proper amount of minor load prior to application of major load during the Rockwell test is highly essential for the following reasons:

i. It minimizes the amount of surface preparation.

ii. It reduces the tendency for ridging or sinking of the test materials around the indenter.

iii. As the indenter is imbedded to the sub-surface by the minor load, the entire effect of surface finish is eliminated.

4.6.7. Precautions to be taken during Rockwell Hardness Test:

Rockwell hardness test is very useful and reproducible if the following precautions are taken. The precautions listed below also apply equally to other hardness tests.

1. The indenter and the anvil should be clean and rigidly seated.

2. The test surface should be clean, dry smooth and free from oxides. A rough surface is usually adequate for Rockwell test.

3. The surface should be flat and perpendicular to the indenter.

4. The thickness of the specimen should be such that bulge is not produced on the reverse side of the test piece. The minimum thickness of the specimen should be 10 times the depth of indentation. Test must be made only on single thickness of the material.

5. Spacing between the indentations should be 3-5 times the diameter of the indentation.

6. Speed of loading should be standardized.

7. The Rockwell hardness value will be lower on the convex surface than on a flat test piece of the same material. For a concave surface or internal diameter the Rockwell hardness value will be higher. Hence, V-blocks should be used to get error free reading.

4.7. Vickers Hardness Test:

Vickers hardness test is another indentation hardness test where a square-based diamond pyramid having an angle 136° between the opposite faces at the vertex is impressed into the surface of the test piece. Both the diagonals of the indentation are measured after removal of the load to calculate the harness value. The geometry of indenter and indentation is shown schematically in the fig.4.6. An included angle of 136° between the opposite faces is chosen because it approximates the most desirable ratio of indentation diameter, d to the indenter diameter, D of the Brinell hardness test.

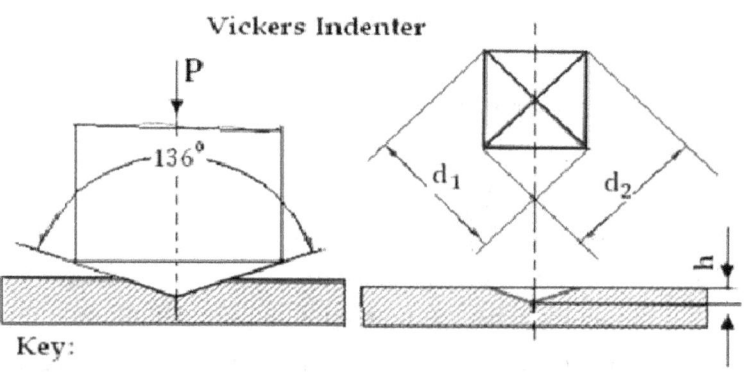

Vickers Indenter

Key:

P = Test load ,in kg.
d_1 and d_2 = Length of the diagonals, in mm
h = Depth of the indentation, in mm

Fig.4.6. Geometry of indenter and indentation in a VHN Test

However, for the diamond pyramid indenter, a *d/D* ratio of 0.375(average of 0.25 and 0.50) is chosen which results in an apex angle of 136° of the conical indenter, as a result the Vickers Hardness Number, VHN is nearly identical to the BHN so long as Brinell impressions are of normal depth (satisfying the condition, *d/D* ratio between 0.25 to 0.50). Due to the shape of the indenter, this test is frequently called Diamond Pyramid Hardness Test and the hardness number is denoted as DPH.

Vickers hardness number is defined as the load divided by the surface area of the impression made on the test piece by the indenter. The indenter impression left on the specimen surface is a square one under ideal condition. However, both the diagonals d_1 and d_2 as shown in the fig.4.6 are measured and averaged out in terms of millimeter for hardness calculation.

Mathematically:

$$\text{VHN or DPH} = \frac{2P\sin(\theta/2)}{d^2} = \frac{1.854P}{d^2}$$

where, *P* is the load applied which usually varies from 1- 120kg, *d* is the average of the diagonals($\frac{d_1+d_2}{2}$) in mm and $\theta = 136°$.

4.7.1. Application of Vickers Hardness Test:

Vickers hardness test has received fairly wide acceptance for research work because it provides a continuous scale of hardness for a given load starting from a very soft material with VHN 5 to extremely hard material with VHN 1500. Excepting for very high loads, the VHN is independent of the load applied.

4.7.2. Designation of VHN:

Example: 550 HV indicates a Vickers hardness value of 550 with a test load of 120kg applied for 20 seconds.

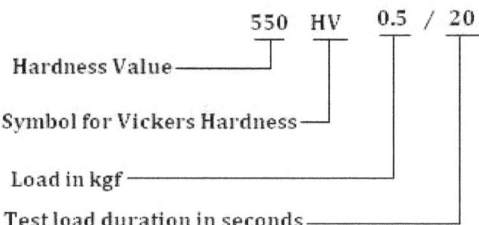

4.7.3. Test Procedure:

I. Select the appropriate test load.

II. The test piece to be placed tightly so that displacement dose not occurs during test. The surface of the test piece should be brought into focus under the objective lens and an area should be selected for indentation.

III. Then the indenter is brought into contact with the test surface and test forces applied for 10-15 seconds perpendicular to the test surface.

IV. Test force is removed and both the diagonals of the indentation are measured. Average of the diagonals (d) is calculated out. The Vickers hardness value is evaluated with the mathematical formula: $VHN = \dfrac{2P \sin(\theta/2)}{d^2} = \dfrac{1.854P}{d^2}$.

The terms used in the formula have been explained earlier.

4.7.4. Limitation of Vickers Hardness Test:

1. This test is not accepted widely for routine work as it is slow.
2. The test requires extensive surface preparation.
3. High degree of skill is required to perform this test thus allows greater change due to personal errors.
4. In place of a perfectly square indention sometimes pin-cushioned or barreled indentations may be formed reflecting either lower or higher hardness values erroneously. Pin-cushioned indentation is a result of sinking-in of the material around the flat faces of the indentation. This condition is observed in extremely soft or annealed materials. This results in overestimation of the dimension of the diagonals and predicts a lower hardness of the material under the test than the actual. Barrel shaped indentation is found in the cold worked materials which results from ridging of material around the faces of the indenter. The diagonal measurement in this case is low and the hardness calculation is erroneously high. Both barreled and pin-cushioned indentations are shown in the fig.4.7.

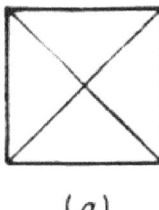  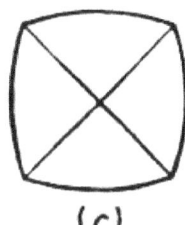

(a)   (b)   (c)

Fig.4.7.(a) Perfect; (b) Barreled; (c) Pin-Cushioned indentations.

4.8. Knoop Hardness Test:

In principle Knoop hardness test is an indentation hardness test. It uses a rhombic-based diamond pyramid indenter having an included longitudinal angle of 172.5° and an transverse edge angle of 130° as shown in the fig.4.8. The indenter is forced into the surface of the test piece with load of P. Then the load is removed and the length of the long diagonal is measured.

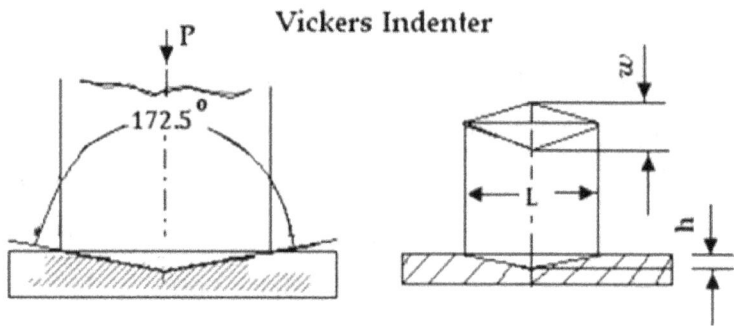

Key: P = Test load in kg.
L = Length of the long diagonal of indentation, in mm
w = Length of the short diagonal, in mm.
h = Depth of the indentation. in mm

Fig.4.8. Indenter geometry of Knoop Hardness Tester.

Knoop Hardness Number, KHN or HK $= \dfrac{P}{L^2 C}$

where P is the Load applied in Kg.
L is the Length of the long diagonal of indentation in mm.
C is the Indenter constant = 0.07028 which relates to the projected area of indentation to the square of the long diagonal (L).

4.8.1. Designation of Knoop Hardness:

Example: 550HK 0.5/20 indicates Knoop Hardness value of 550 determined with a test force of 0.5 kgf acting for 20 seconds on the specimen surface.

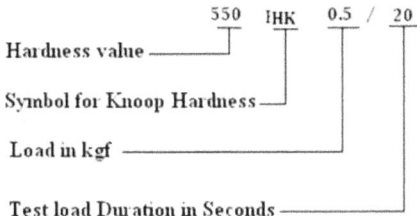

Hardness value ⎯⎯⎯⎯⎯⎯⎯⎯⎯⎯⎯
Symbol for Knoop Hardness ⎯⎯⎯
Load in kgf ⎯⎯⎯⎯⎯⎯⎯⎯⎯⎯⎯⎯
Test load Duration in Seconds ⎯⎯

4.8.2. Test Conditions:

I. Load:

The load selected should be compatible with the hardness & thickness of the test piece. If scope exists between different loads, the heaviest load should be used.

II. Surface:

Top & bottom surfaces should be flat, smooth and parallel. The test surface should be fine polished to obtain a well defined indentation.

III. Thickness:

The thickness of the test piece should be at least 3 times the length of the long diagonal. After the test no bulge should appear on the opposite bottom surface

IV. Mounting of the Test Piece:

Test pieces of small cross-section or irregular shape may be mounted using hot or cold stetting resin to facilitate proper surface preparation.

V. Spacing between Indentations:

a. Spacing between the edges and the indentation marks:

(i) Should at least be: 2.5 times the length of shorter diagonal of indentation incase of steel, copper & alloys and

(ii) 3times the length of shorter diagonal of indentation in case of light metals like lead, tin, aluminium & their alloys.

b. The distance between the centers of two adjacent impressions:

(i) Should at least be 3 times the length of shorter diagonal of indentation in case of steel, copper and copper alloys and

(ii) 6 times the length of shorter diagonal of indentation in case of soft alloys like lead, tin, aluminium and their alloys.

4.9. Interrelation between Various Hardness Tests:

Now standard tables are available to convert one hardness value to hardness value. This helps in comparison work. There have been several empirical relationships to convert the result of one particular test to another type. Since hardness test does not measure a well defined property of the material and since all the tests make different measurements, there exist no universal hardness-conversion relationship. Hence it is important to realize that hardness conversions are only empirical in nature. The most important relationship exists for steels which are harder than 240 BHN.

I. BHN - VHN Conversion:

BHN and VHN values are almost identical upto 400BHN. Beyond 400HB the hardened steel ball indenter used for testing starts deforming and records a lower hardness value while the diamond indenter used in Vicker tester remains undeformed and records a higher hardness value for the same material being tested.

II. BHN-Rockwell Conversion:

i. $BHN = \dfrac{7300}{130 - R_B}$ for $R_B\ 35 - 100$

ii. $BHN = \dfrac{14,20,000}{(100 - R_C)^2}$ for $R_C\ 20 - 40$

iii. $BHN = \dfrac{25,000}{100 - R_C}$ for $R_C\ 40 - 70$

4.10. Interrelation between Hardness and Strength:

It is important to mark that both BHN and VHN have the same unit of kg/mm² which is also same for the tensile strength of the material. As both the tests measure the force required for indentation(plastic deformation), there exists relationship for converting the hardness values to ultimate tensile strength of the materials. This conversion is helpful in design and quality control as it helps in rapid determination of such parameter without actual testing. However, these conversions are applicable only to carbon steels upto 1.0 wt% carbon in a fully annealed or tempered condition. The relationships between ultimate tensile srtength and BHN are as follows:

i. UTS for annealed steels is: $0.35 \times (BHN)$ kg/mm²

ii. UTS for quenched & tempered steel: $0.324\ (BHN)$ kg/mm².

*****

Chapter 5
FATIGUE TEST
5.1. Introduction:

It has been recognized since 1830 that a metallic component subjected to a repetitive or fluctuating stress will fail at a stress much lower than that is required to cause its failure under uniaxial tensile test. The failure occurring under the condition of fluctuating load is called fatigue failure. Fatigue failures are generally observed only after considerable period of service life. Fatigue has become a very important mode of failure for automotive, aircraft, compressor pump and turbine components where the components are subjected to repeated fluctuating load and vibration. It is generally observed that at least 90% of all service failures are due to fatigue and other mechanical reasons.

Fatigue failure in particular is very critical because it occurs without any prior warning. It results in a brittle fracture without any gross deformation at the failure site. On a macroscopic scale, the fracture surface is normal to the direction of principal tensile stress acting on the component. Usually this failure can be recognized by observing the appearance of the fracture surfaces. The surfaces appear smooth due to repeated rubbing action against each other as the crack propagates through the material. On microexamination, the progress of the fracture is indicated by series of rings or sea beach sand profile as produced by the water waves. These beach marks are in the progressive direction of the fatigue crack. A rough region is seen towards the end of the failure surfaces as the remaining cross-section becomes no longer able to carry the applied load due to propagation of fatigue crack.in the metallic object.

5.2. Factors Responsible for Fatigue Failure:

The basic factors responsible for fatigue failure are:

i. A minimum tensile stress of sufficiently high value must act.
ii. There must be a large fluctuation in the applied stress.
iii. A sufficiently large number stress cycles must be applied.
iv. Including the above three major factors, host of other variables such as stress concentration, corrosion, operational temperature, over loading, residual stress, combined stresses and metallurgical factors also aid the fatigue failure.

A complete knowledge regarding fatigue failure is yet to be obtained even today and the above factors are only related to the fatigue failure empirically.

5.3. Alternating or Fluctuating Stress Cycles:

All the engineering components are subjected to stress during their service life. However, the stress condition may vary widely. The stress acting may be static or dynamic. Under the dynamic condition the stress fluctuates repeatedly either in a regular or irregular fashion. General types of stress cycles which may cause fatigue are defined schematically in the fig5.1(a) to (d).

(i) Completely reversed sinusoidal form of stress cycle:

This type of stress cycle has a defined maximum and minimum stresses of equal amplitude and the cycle is repeated with definite time interval (fig.5.1a.).

(ii & iii). Completely reversed sinusoidal form of stress cycle of unequal amplitude:

This type of stress cycle is repeated in regular time interval but the maximum and minimum levels of the stresses are unequal as shown in the fig.5.1b & c.

iv Complex Stress Cycle:

This kind of stress cycle(fig.5.1d). is the most complex and critical as nothing can be predicted about the amplitude of the stress variation acting on the component. Further there is no regularity in the stress cycle with respect to time. This type of stress cycle is observed in the aircraft wings & automobile parts where the components are subjected to unpredictable overloads due to various reasons during their service life.

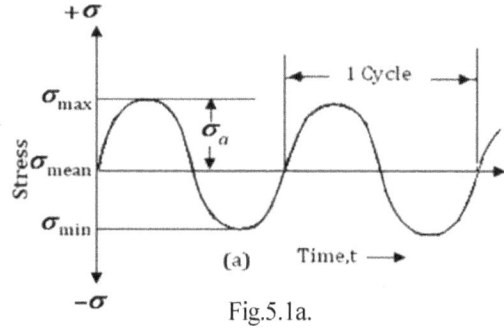

Fig.5.1a.

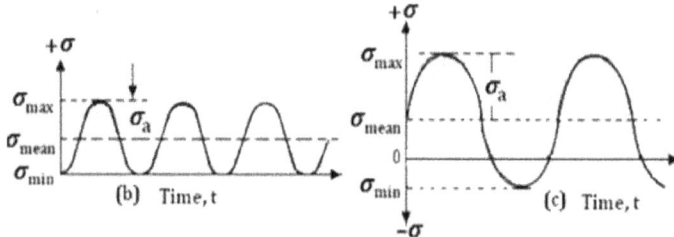

Fig.5.1b & c.

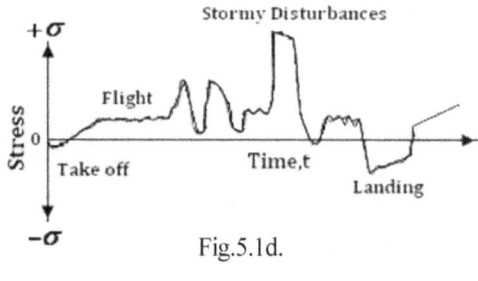

Fig.5.1d.

5.4. Terms Related to the Stress Cycles:

A fluctuating stress cycle can be considered be made up of two components:

i. A mean or steady stress

ii. An alternating or variable stress.

The range of the stress cycle is also important. Mathematically,

$$\sigma_{mean} = \frac{\sigma_{max} + \sigma_{min}}{2}$$

$$\sigma_{amplitude} = \frac{\sigma_{max} - \sigma_{min}}{2}$$

$$\sigma_{range} = \sigma_{max} - \sigma_{min}$$

Stress Ratio, $SR = \sigma_{min}/\sigma_{max}$ and

Amplitude Ratio, $AR = \sigma_{range}/\sigma_{mean} = (1-R)/(1+R)$

5.5. Presentation of Fatigue Data:

The basic method of presenting fatigue data is by means of a *S-N* curve which is a plot of stress, *S* versus the number of stress cycless, *N* to failure. A log-scale is usually used to represent the number of cycles as it is quite high for the failure to occur. The stress values are usually nominal stress without any adjustment for stress concentration. Determination of fatigue properties is made under completely reversed bending condition as this condition is the most severe one for most of the materials.

Step 1: To plot the *S-N* curve, large number of specimens are prepared from the material of same stock and marked serially. Usually a long rod is used for the purpose.

Step 2: The first specimen is tested at a high stress where the failure is expected in a fairly less number of stress cycles. The number of stress cycles the specimen withstands before fracturing is noted.

Step 3: Then the stress level is decreased for each succeeding test specimen until one or two specimens do not fail within the specified number of stress cycles.

Step 4: Finally the stress, σ is plotted against the log N for each specimen, where N is the number of stress cycles to failure of the specimen. The plot of the S-N curves for different materials are shown schematically in the fig. 5.2.

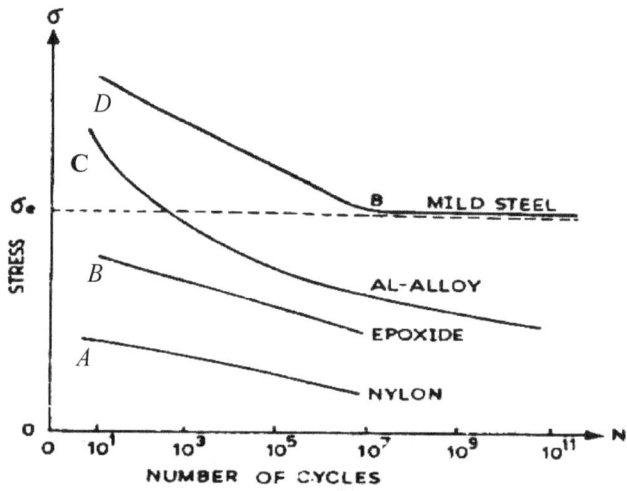

Fig. 5.2. *S-N* curves for different Materials

Of the four fatigue curves shown above schematically, *A* represents low carbon steels, *B* represents aluminium alloy, *C* represents epoxy resin & *D* represents nylon.

5.1. Analysis of the Fatigue Curve:

i. From the shape of the curves it can be seen that the fatigue life is the number of stress cycles the material can sustain or endure before failure which increases with decreasing stress level and vice versa.

ii. The *S-N* curve for mild steel shows a distinct feature of becoming parallel to *N* axis below a certain limiting stress value. This signifies that below this limiting stress value the material can withstand infinite number of stress cycles without fracturing.

The critical value of the stress at which the materials sustain infinite number of stress cycles in known as *Fatigue Strength* or *Endurance Limit*. Curve *A* is typical of carbon steels and titanium alloys where the fatigue limit is clearly reflected by the shape of the curve. For the curve *B*, which is typical of non-ferrous metals & alloys, stress drops continuously with respect to *N* and shows no definite fatigue limit as in the case of carbon steels. Hence, these materials do not possess a true fatigue limit. For such materials, it is a common practice to define an endurance limit as the maximum stress at which the materials sustain an infinite number of stress cycles before failure. This number is arbitrarily taken as either 10^8 or 5×10^8 stress cycles for all practical purposes.

5.7. Determination of Fatigue and Endurance Limit:

The method of estimating fatigue or *endurance limit* of the materials is to draw the *S - N* curve and then draw a line on the *S - N* curve parallel to stress(*S*) axis corresponding to 10^7 cycles on the *N* axis for ferrous materials and 5×10^8 cycles for non-ferrous alloys to intersect the *S - N* curve. Then drop a perpendicular from the point of intersection to the stress axis to measure the fatigue stress or endurance limit. In fig.5.2 the endurance limits of carbon steel and aluminium alloy are shown schematically. It should be noted that both fatigue and endurance limits of mild steel are same.

5.8. Types of Fatigue Tests on the basis of Stress Cycles:

Fatigue tests can be classified into:

a. High Cycle Fatigue.

b. Low Cycle Fatigue.

If the fatigue test is conducted upto 10^8 cycles or more then it is termed as high cycle fatigue and if the test is conducted upto 10^5 cycles or less it is termed as low cycle fatigue. The high cycle fatigue test is conducted with specified load or stress while the low cycle fatigue is conducted with controlled or specified strain.

5.8.2. Low-Cycle Fatigue:

The usual way of presenting low cycle fatigue test results is to plot plastic strain range, $\Delta\varepsilon_p$ against number of cycles. A straight line is obtained when $\Delta\varepsilon_p$ and N are plotted on a log-log scale. This type of behaviour is known as Coffin-Manson relation. Mathematically, Coffin-Manson relationship is : $\Delta\varepsilon_p = \varepsilon_f'(2N)^c$, where, $\Delta\varepsilon_p/2$ is the plastic strain amplitude, ε_f', is the fatigue ductility coefficient defined at strain intercept $2N=1$, $2N$ is the number of stress reversals to failure(N cycles) which is approximately equal to the true fracture strain ε_f for many metals, c is the fatigue ductility exponent(-0.5 to -0.7 for most metals).

5.8.2. Basquin's Equation:

The S-N curve in high cycle region is sometimes described by Basquin equation: $N \times (\sigma_a)^p = C$, where, N is the number of stress cycles, σ_a is the amplitude of stress, p &C are empirical constants.

5.9. Theory of Fatigue Failure:

No theory has yet been developed to explain the fatigue failure comprehensively and conclusively. However, the failure is most probably due to the build- up of the slip lines in a particular region. The entire process of fatigue failure is conveniently classified into three (3) different stages:

i. Slip Band Crack Growth:

This stage involves deepening of the initial cracks developed on the planes of high shear stress and is termed StageI crack growth.

ii. Crack Growth on Planes of High Tensile Stress:

This stage involves growth of well defined cracks produced during the StageI in a direction normal to the maximum tensile stress. This is usually called StageII crack growth.

iii. Ultimate Ductile Failure:

This is the last stage of the fatigue failure which occurs when the crack dimension is sufficiently largely and the remaining cross section of the specimen cannot support the applied.

5.10. Factors Affecting the Fatigue Property of Metals:

Following factors greatly affect the fatigue property of the metals & alloys:

i. Composition.
ii. Stress concentration.
iii. Size effect of the component.
iv. Surface condition.
v. Residual stress.

5.10.1. Composition:

Fatigue strength of a metallic material is not directly influenced by the alloying elements. However, those alloying elements which increase the tensile strength are also found to increase the fatigue strength. In case of steels, carbon is found to have maximum effect on fatigue strength. It is established that the alloys which undergo strain hardening exhibit a sharp knee on the *S-N* curve. In case of steels the knee is found to decrease and the curve flattens as the carbon and nitrogen content of the steel is reduced.

5.10.2. Stress Concentration:

Fatigue strength is seriously reduced by the introducing stress raisers, such as notches, holes or key ways into the components. The stress raiser points are found to initiate fatigue cracks. All engineering components contain stress raisers, hence during design, sharp corners, notches or keyholes should be avoided as for as practicable.

5.10.3. Size Effect:

To predict the fatigue performance of a large machine member from the laboratory test results conducted on small test specimen is always associated with inherent practical problems. The major reason being the the number of defects present in the large component will far exceed the number of similar defects present in the smaller test specimen of the same material. Due to the above fact, in most cases, size effect exists. The fatigue life of large machine members will obviously be lower than the predicted value on the basis of laboratory test results.

5.10.4. Surface Condition:

Practically all fatigue failures start at the surface. There is ample evidence that surface conditions have pronounced effect on the fatigue life of a machine component. The factors which affect the fatigue strength due to surface conditions are:

i. Surface roughness or stress raiser on the surface.
ii. Change in the fatigue strength of the material at the surface.
iii. Change in the residual stress on the surface of the material.

A residual tensile stress at the surface greatly hampers the fatigue strength as it helps in opening up the existing microcracks on the surface. However, residual compressive stress at the surface greatly improves the fatigue strength as it always helps in closing the existing micro cracks on the surface. A component with a rough surface finish will have a lower fatigue strength compared to the same component with a finer source finish. Other conditions which affect fatigue strength or life are oxidation & corrosion on the surface.

5.10.5. Residual Stress:

A residual compressive stress on the surface of the component greatly improves the fatigue strength of the material. Compressive stress can be developed at the surface level by:

I. Shot Peening or Shot Blasting.
II. Surface Rolling using Contoured Rolls.

5.11. Metallurgical Factors Affecting Fatigue Strength:

Metallurgical factors which affect the tensile strength of the metals and alloys will in general affect the fatigue strength.

The general factors are:

i Solid solution alloying increases the fatigue strength.

ii. Finer the grain size higher is the fatigue strength.

iii. In case of case hardened steel the fatigue strength decreases as the percentage of martensite formed decreases.

iv. Tempering after quench hardening increases the fatigue strength.

v. Even a trace of decarburization at the surface adversely affects the fatigue strength of the steel. Unless otherwise mentioned, the fatigue strength for steels is taken to be half of its tensile strength.

5.12. Temperature and Fatigue Strength:

Fatigue strength increases with decrease in temperature at which test is conducted. Unlike impact strength, there is no sudden fatigue embrittlerment with the fall of temperature.

5.13. Thermal Fatigue:

Metals & alloys expand & contract on heating and cooling respectively. This alternative expansion & contraction due to change in temperature develops stresses and may cause fatigue failure. This explains that the stresses which produce fatigue failure at high temperature do not necessarily come from mechanical sources. The fatigue failure may also be due to fluctuating thermal stresses. Thermal stresses are caused when the change in dimensions of a member, due to change in temperature, are prevented by some kind of constraint. For example, a metallic bar fixed at both the ends will develop a thermal stress, due to a temperature differential of ΔT. Mathematically, thermal stress produced is expressed as:

$\sigma_T = \alpha \times E \times \Delta T$, where

σ_T = Thermal stress.

ΔT = Fluctuation in temperature or temperature differential,

α = Linear coefficient of thermal expansion

E = Elastic modulus of the material under test.

If failure occurs by a single application of thermal stress, the condition is called *Thermal Shock*. while the failure due to repeated application of thermal stress of lower magnitude is called *Thermal Fatigue*. The later condition is frequently present in high temperature equipments. Austenitic stainless steels are in particular sensitive to thermal fatigue because of their low thermal conductivities and high coefficient of thermal expansions. Thermal fatigue is related to the parameter, $(\sigma_f \times k)/E\alpha$, where, σ_f is the fatigue strength, k is the thermal conductivity, α is the coefficient of linear thermal expansion & E is the elastic modulus. A higher value of $\sigma_f \times k/E\alpha$ reflects a better resistance to thermal fatigue.

5.14. Fatigue Testing Machine:

The fatigue limit of a material is determined in a Rotating Beam Fatigue Testing Machine developed by Moore. A polished specimen (beam) is placed as a simple supported beam and loaded in pure bending by a dead weight, W. The beam is rotated in the ball bearing at speed ranging from 3,000 to 10,000 rpm. A variable speed or a constant speed motor with a gear box is used for this purpose. There are number of tests to measure the fatigue limit or strength of a material. A relatively simple alternating stress obtained by reversed bending in a rotating beam fatigue testing machine is generally used for its relative simplicity.

The fig.5.3. shows the basic components of the fatigue testing machine schematically. However, fatigue tests are statistical in nature. One of the major elements of the fatigue testing machine is the high speed electric motor rotating at 10,000 rpm. Next to the motor is a large bearing. The purpose of the bearing is to relieve the motor of the large bending moment which is applied to the specimen. The specimen is mounted in the collects that serve as grips. The application of the force produces bending in the specimen, so that upper surface is under tension while the lower surface is under compression. When the specimen rotates by the action of the motor, any given position on the surface of the specimen, alternates between a state of, maximum tensile and compressive stresses. The top most layer of the rotating beam specimen always remains in compression while the bottom layer remains in tension. Thus a sinusoidal stress is produced on the surface of the specimen during each rotation. A mechanical revolution counter records the number of stress reversals the specimen undergoes upto failure. The machine stops when the specimen breaks.

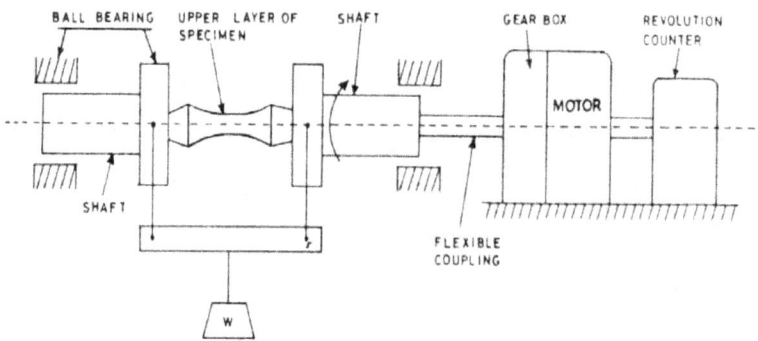

Fig.5.3. Moore Rotating Beam Fatigue Testing Machine

5.15. Empirical Relations to find Endurance Limit:
I. Gerber's Parabolic Relation for Ductile Materials:

$$\sigma_a = \sigma_e [\frac{1}{f_{os}} - (\frac{\sigma_m}{\sigma_u})^2 f_{os}].$$

σ_u = Ultimate tensile or compressive stress of the material.

f_{os} = Factor of Safety.

σ_m = Mean Stress = $\dfrac{\sigma_{max} - \sigma_{min}}{2}$.

σ_e = Endurance limit of the Material.

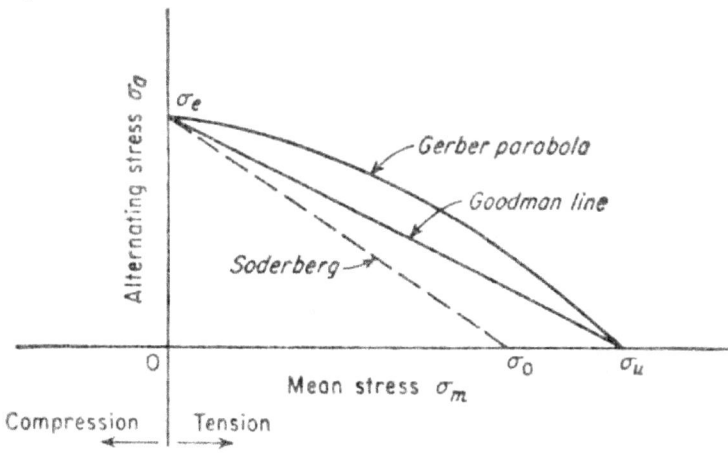

Fig.5.4. Alternative Method of plotting Goodman Diagram

II. Goodman's Straight Line Relation:

This relation is valid for both ductile and brittle materials.

$$\sigma_a = \sigma_e [\frac{1}{f_{os}} - \frac{\sigma_m}{\sigma_u}],$$ where related terms are as explained above.

III. Soderberg's Straight Line Relation:

$$\sigma_a = \sigma_e [\frac{1}{f_{os}} - \frac{\sigma_m}{\sigma_y}],$$ where σ_y is the yield strength of the material.

5.16. Philosophy of Fatigue Design:

There are several philosophies of fatigue life design as:

Infinity Life time:

This design criterion is such that the working stress on the component is some fraction of the fatigue limit of the steel. Hence the design is for infinite cycles of uniform alternating stress. This is the oldest fatigue design philosophy.

Self –Life Design:

Self life design is based on the assumptions that the part is initially flawless and has an infinite life at constant stress. Further it is develops cracks during service life due to large amount of stress scatter. For example aircraft and automobile components are designed with a factor safety of around 4. The factor of safety accounts for the environmental effects and etc. Bearing are rated by specifying the load at which 90% of all the bearing are expected to withstand the given life time.

Fail Safe Design:

In fail safe design, the condition is such that the fatigue crack will be detected & repaired before any failure takes place.

Damage Tolerant Design:

This is the latest design philosophy, merely an extension of self-life design. In damage-tolerant design it assumed that fatigue cracks exist in the material prior to use. Fracture mechanics is applied to determine whether the crack will grow large enough to cause failure or not before they are detected by periodic NDT inspection.

****

Chapter 6

CREEP TEST

6.1. Introduction:

It is important to note that strength of the metals and alloys decreases with increase in temperature. As temperature increases the mobility of atoms increases rapidly with creation new slip systems. Hence, deformation at grain boundaries becomes an added possibility with increased probability of dislocation movement. The entire phenomena are related to diffusion. Hence, high temperature mechanical properties of metals & alloys are significantly affected by all difussion controlled processes in the material.

The principal use of high temperature material have been in the field of steam power plants, oil refineries and chemical plants with a minimum operating temperature of 500ºC. Further, with the advent of gas turbines the operating temperature has gone upto 800ºC. Similarly rocket engine components & ballistic missile nose cones pose much greater problem of designing new high temperature materials. The most important characteristic of the high temperature strength is that it must always be considered with respects to some time scale. The tensile properties of metals & alloys are independent of time at room temperature. However, at elevated temperatures strength of the materials becomes dependant both on strain rate and time of exposure to higher temperature. Hence, metals & alloys subjected to elevated temperature will undergo time dependant deformation. Higher temperature cannot be defined universally. A particular temperature which is higher for a particular material may not be higher for another due to the differences in their melting points. To take care of this anomaly, temperature is often expressed as *Homologous Temperature*.

Homologous Temperature is defined as the ratio of test temperature to the melting temperature of that material expressed in kelvin scale. Hence, for comparing the high temperature deformation of two different materials the test should be carried out at equivalent homologous temperature. The time dependant deformation of the material becomes significant at homologous temperature greater than 0.5.

6.2. Creep:

Permanent deformation of a material under steady load or stress as a function of time which is defined as **Creep**. It is a thermally actuated process hence is influenced by test temperature. Creep occurs at room temperature in materials like lead, zinc and etc. Consideration and study of creep deformation is extremely important in applications like:

1. Industrial Belts.
2. Gas Turbine Blades.
3. I.C. Engine Piston.
4. Rockets & Missiles.
5. Heat Exchanger Tubes and etc.

6.3. Elevated Temperature Tests:

The creep behaviour of the materials stressed at elevated temperature depends both on test duration and the stress level. As the life expectancy of the machine part is usually high it may not be possible to run a test for many years to evaluate the required property. In fact, the test is carried out for a much shorter period and the test result is extrapolated to provide useful data.

Test is essential to determine the dependency of creep deformation & rupture strength on exposure time at a predefined stress level. Important creep properties used widely in designing of components for various applications are:

1. Evaluation of creep deformation strength or yield strength.
2. Evaluation of creep rupture strength or ultimate strength.

Creep deformation strength is defined as the highest stress that a material can withstand for a specified time period at a certain predefined temperature without excessive deformation. Similarly creep rupture strength is defined as the highest stress that a material can withstand for a specified time period at certain temperature without rupture. The above strengths are generally calculated to determine the life expectancy of the components at their operating temperatures. Creep Test measures the dimensional changes resulting from exposure to elevated temperature at constant load or stress while the *Stress Rupture Test* measures the effect of temperature on long time load bearing characteristic of the materials.

6.4. Creep Test:

As progressive plastic deformation a constant stress is called creep this test is simply a tension test run at constant load and temperature. During the test the specimen is places inside a furnace and heated for few hours. Then it is subjected to constant load by a lever and dead weight system. In due course of time, creep deformation starts in the specimen which is recorded at certain intervals of time. Marten's optical extensometer records the strain of the specimen with an accuracy of 0.001 mm. The total creep test may take few hours to few years as per importance and need.

The stages of deformation is as shown in fig.6.1.

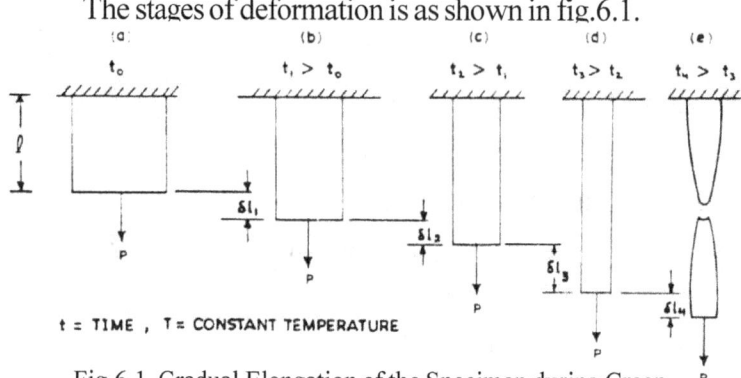

Fig.6.1. Gradual Elongation of the Specimen during Creep

The total creep or elongation or strain, ε is measured for the entire duration of the test. The plot of strain, ε versus time, t is known as creep curve and is of great engineering importance. This curve is utilized to measure the minimum creep rate.

6.5. Idealized Creep Curve:

The curve A in the figure below represents an idealized creep curve. After rapid initial elongation, the deformation rate decreases with time and then reaches a steady state. Finally deformation rate increases rapidly with time till the specimen fractures. The creep rate is slop of the curve during steay state creep.

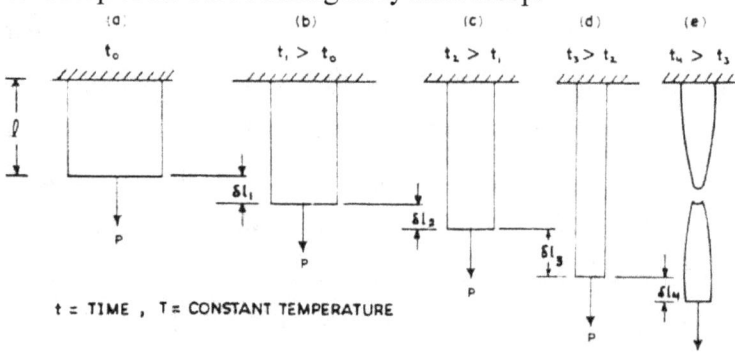

The stages of deformation is as shown in fig.6.1

The creep curve in general shows three distinct stages as shown in the fig.6.2:

Stage I: Primary Creep,

Stage II: Secondary Creep or steay state Creep and

Stage III: Tertiary Creep.

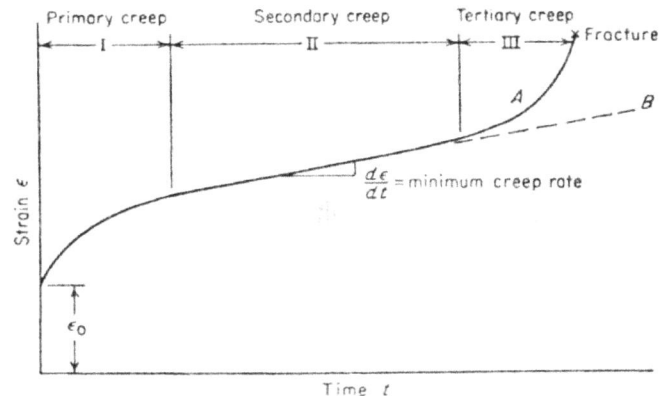

Fig.6.2. Typical Creep Curve showing Three Stages of Creep

6.5.1. Stage I:

The 1st stage of creep is known as primary or transient creep that shows a decreasing creep rate. During this stage the creep resistance of the material increases by virtue of its own deformation and simultaneous work hardening. At low temperature and stress, Stage I creep is significant while it vanishes as temperature & stress increases.

6.5.2. Stage II:

Stage II creep is known as secondary creep. This is the period of nearly constant creep rate and is of great engineering importance.

This constant creep rate results from the balance between the two competing processes that is strain hardening & recovery. For this reason secondary creep which has the lowest or minimum creep rate known as **Steady State Creep**.

6.5.3. StageIII:

Stage III of the creep curve is known as **Tertiary Creep**. Tertiary creep occurs when there an effective reduction in cross-sectional area of the specimen either due to necking or due to formation of internal voids. Tertiary creep occurs at high stresses or at high temperatures leading to ultimate failure of the material.

6.6. Steady State Creep: An Parameter in Engg. Design:

The minimum creep rate is the most important design parameter derived from the creep curve from this minimum creep rate an average estimate regarding the life span of the machine component can be made. They are two standard methods of measuring this parameter. They are:

i. The stress required to produce a creep rate of 0.0001% per hour.

ii. The stress required to produce a creep rate of 0.000001% per hour or 1% per 100,000hrs. The first criterion is more typical of jet engine components while the second one is important for steam turbine or similar components. A log-log plot of stress versus minimum creep rate frequently results in a straight line useful in design engineering components.

6.6. Andrade's Analysis of Creep Curve and Creep Laws:

Andrade was the first to make a detailed analysis of the creep curve as shown in the fig.6.3. .

According to him the constant stress creep is a superimposition of two separate creep processes which occur after the sudden strain, ε_o.

Stage I Creep Laws:

The StageI of the creep curve is known a transient creep or cold creep. This occurs even at low temperature. Hence, known as primary or cold creep.

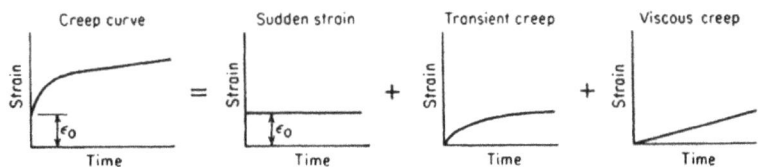

Fig.6.3. Andrade's analysis of the creep curve

i. Parabolic Law:

Transient creep for metals and alloys, $\varepsilon_{cr} = Ct^n$, where C is a constant, n is a power index whose value is 1/3 and t is the time perod.

ii. Logarithmic Law:

Transient creep for plastics and rubber, $\varepsilon_{cr} = k\ln[1+\dfrac{t}{t_i}]$, where k is a constant and t_i is any arbitrary chosen time.

iii. Hyperbolic law:

$\varepsilon_{cr} = \lambda t/(n+1)$, where λ is a constant, n is creep-time constant and ε_{cr} is the creep strain.

Stage II Secondary Creep Law:

This stage of creep is known as secondary or steady state or viscous creep resulting from a balance between the competing processes of strain hardening and recovery. It is important in design.

This stage of creep is also known as hot creep. The law is represented as: $\varepsilon_{cr} = \varepsilon_1 + v_{cr}t$ where, ε_1 is the strain intercept, v_{cr} is the minimum or viscous creep rate. Further, $v_{cr} = A\sigma^n$ for $n > 1$, where A and n are constants.

Stage III or Tertiary Creep:

Third stage of creep is known as tertiary creep, during which creep rate increase rapidly and finally results in failure of the material.

6.7. Mechanism of Creep:

Occurrence of creep in materials is supposed to be the effects of following phenomena:
1. Vacancy Diffusion.
2. Edge Dislocation climb up or climb-down.
3. Grain Boundary Sliding.
4. Screw Dislocations cross-slip.
5. Elastic after effect.

6.8. Factors Affecting Creep:

Both stress and temperature affect the creep.

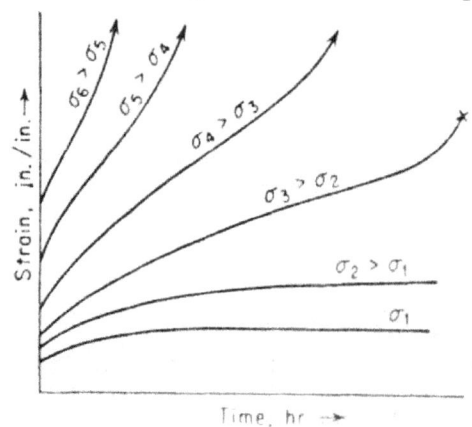

Fig.6.4.i. Effect of Stress on Creep Curve

ii. Effect of Temperature on Creep Curve:

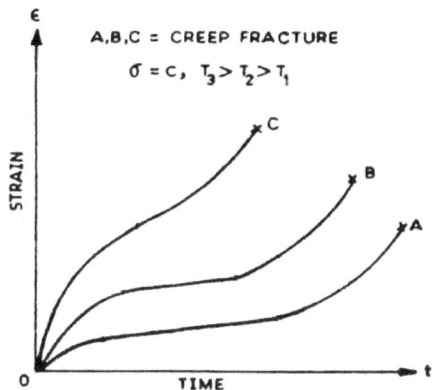

Fig.6.4.ii. Effect of Temperature on Creep Curve

6.9. Creep Resisting Materials:

Most of the high temperature materials are creep resting materials. The effective materials are:

1. Refractories.
2. Tungsten based alloys.
3. Nickel based alloys.
4. Cobalt based alloys.
5. Steel based alloys.
6. Monocrystals of titanium.
7. Thoria (ThO_2) dispersed nickel.
8. Sialon (Si_3N_4 & Al_2O_3) for turbine blades.

Metals with low stacking fault energy have higher creep resistance.

6.10. Stress Rapture Test:

The stress rapture test is basically similar to creep test. However, this test is always conducted at much higher loads and always continued upto the failure. The creep rate in stress rupture test is always higher compared to creep test.

During the creep test our basic aim is to find out the minimum or steady creep rate and to avoid tertiary creep. Usually the total strain during a creep test is often less than 0.5% while in a stress rapture test the total strain may be as high as 50%. Hence simple strain measuring devices with less complicated set-up can be used for stress- rupture test. Due to higher stress and creep rate of the stress-rapture test causes structural changes in the metals at a shorter time. Therefore, the stress-rapture tests are usually terminated in 1000hrs. For the above reasons stress-rapture test has found increased use. The basic information obtained from the stress rupture test is the time taken to cause failure for a given nominal stress at a constant temperature. In this case stress is plotted against the rupture time on a log-log scale. A straight line is usually obtained for each test temperature. Change in the slope of the stress-rupture line is due to structural changes occurring in the material during testing. The structural changes may be oxidation, recrystallization, grain growth, spheroidizing, graphitization, transgranular or intergranular fracture.

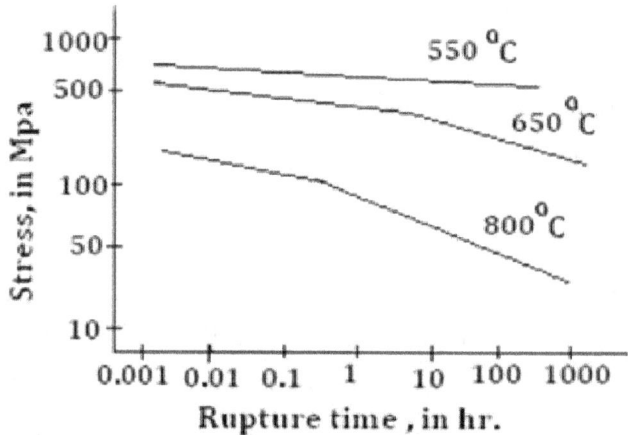

Fig.6.5. Schematic Plot of Stress-Rupture data

6.11. Superplasticity:

Superplasticity is the ability of a material to withstand large deformation in tension without necking. An elongation in excess of 1000% is usually observed in superplasticity. Superplastic behaviour occurs at a temperature, $T > 0.5T_m$, where materials show large extensibility without fracture. At low strain rates the flow stress is also low, so complex shapes may be formed readily under superplastic conditions. The requirements for a material to exhibit superplasticity are:

i. Very fine grained structure where the grain size is less than $10\,\mu m$.
ii. Presence of second phase materials with similar strength which inhibits grain growth at elevated temperature. Most of the superplastic materials are either eutectic or eutectoid alloys. The predominant mechanism for superplastic deformation is the grain-boundary sliding accommodated by slip.

6.12. Classification of Fracture:

Separation or fragmentation of a solid into two or more pieces under applied stress or load is called fracture. Depending on the type of stress applied, fracture may be termed as:

i. Tensile fracture.
ii. Compressive.
iii. Shear.
iv. Fatigue.
v. Creep.
vi. Cleavage fracture and etc.

Further, the mechanism of fracture can be considered to be composed of two distinct phenomena:

i. Crack initiation:
ii. Crack propagation:

Depending upon the extent of plastic deformation prior to failure, the fracture may be classified into two distinct categories:
i. Ductile fracture.
ii. Brittle fracture.

The fracture may be brittle or ductile. But it has been observed that for metallic materials fracture is rarely either purely brittle or purely ductile. Rather most of the fractures are a combination of both.

6.12.1. Brittle Fracture:

Failure of a material with rapid rate of crack propagation and negligible plastic deformation is called brittle fracture. This type of fracture resembles to that of cleavage fracture in ionic solids. Hence, this fracture is also known as cleavage fracture. Broken pieces of material can be assembled back to get the unbroken shape and size of the component. This fracture is often unpredictable as the cracks propagate all of a sudden. The tendency for brittle fracture increase with:
i. Decrease in temperature (subzero temperature).
ii. Increase in strain rate (rate of loading or stressing is rapid).
iii. Triaxial state of stress (a condition usually produced by the notches).

6.12.2. Ductile failure:

In a ductile fracture, a material absorbs larger amount of strain energy prior to its failure compared to brittle fracture. The fact that the material absorbs strain energy prior to failure implies plastic deformation will occur before the fracture takes place.

Ductile fracture may take several forms. HCP single crystal metals may slip on successive basal planes till the crystal separates by shear. Very ductile polycrystalline metals like gold, silver or lead may be drawn up-to a point before failure. During the tensile fracture of moderately ductile metals, the plastic deformation eventually produces a necked region. The fracture begins at the centre of the specimen and extends outward to the surface by a shear mode. This results in a familiar *cup and cone* fracture. So it can only be told that ductile fracture is more acceptable in engineering design as the fracture is associated with large amount of plastic deformation prior to failure which provides prior warning sufficiently before theactual fracture takes place.

6.12.3. Gansamer Classification of Fracture:

Fractures are classified with respect to characteristic such as: strain to fracture, crystallographic mode of fracture and appearance of fracture:

Behaviour Described	Fracture Term Used
Crystallographic Mode	Shear Or Cleavage
Appearance of Fracture	Fibrous Or Granular
Strain To Fracture	Ductile Or Brittle

6.12.4. Typical Tensile Fracture of Metals:

i. Cup and Cone fracture - Mild Steel.
ii. Fibrous fracture   - Wrought iron.
iii. Star fracture   - Lead, Magnesium
iv. Granular fracture   - Cast iron.

The fractured surfaces frequently consist of a mixture of fibrous and granular zones as shown in the fig.7.55.

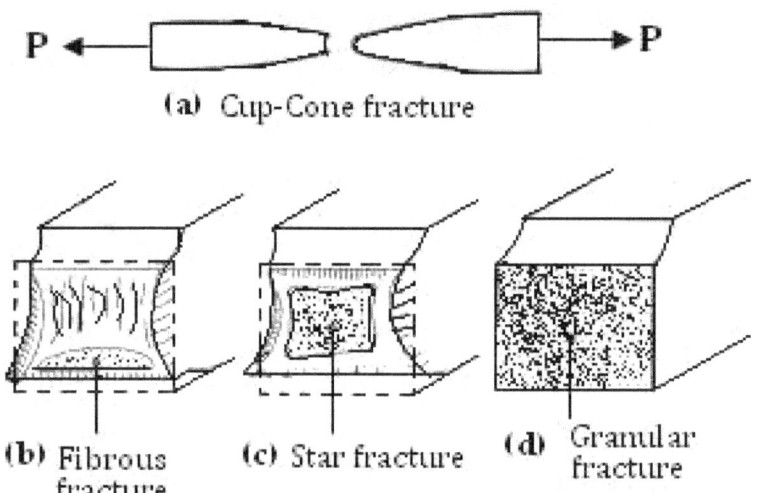

(a) Cup-Cone fracture

(b) Fibrous fracture (c) Star fracture (d) Granular fracture

Fig.6.6. Typical Appearance of Fratured Surfaces

It is customary to report the percentage of the surface area represented by each of these categories. Based on the metallographic examination, fracture in polycrystalline material is also classified as:

i. Transgranular fracture (crack propagating through the grains).

ii. Intergranular fracture (crack propagating along the grain boundaries).

6.12.5. Mechanics of Ductile Failure:

Ductile failure is associated with large plastic deformation prior to failure. Failure progresses in successive stages as follow:

i. As the sample is loaded in tension, a neck forms beyond the ultimate strength of the material.

ii. Crack starts to nucleate at the brittle phase particles such as cementite or oxides present in the material.

iii. With increasing stress, the nucleated crack grows to a size of around one millimeter.

iv. The crack proceeds outwardly towards the surface in a direction perpendicular to the applied stress.

v. The crack propagates in a direction 45° to the tensile axis finally resulting in a cup & cone fracture.

6.12.6. Griffith's Theory of Brittle Fracture:

Materials fail at a much lower stress than its theoretical strength. The first explanation regarding this discrepancy between the theoretical & practical strength of the crystal was proposed by Griffith. This theory in its original form is applicable only to perfectly brittle materials such as glass and ceramics. Griffith proposed that brittle materials contains a population of fine cracks which produces a stress concentration of sufficient magnitude so that theoretical cohesive strength is reached in localized regions at a nominal stress well below the theoretical strength. Though this theory could not be applied directly to the metals, it had a great influence on modeling fracture of metals and alloys.

6.12.7. Griffith's Criterion to Propagate a Crack:

A crack will propagate through the material when the decrease in elastic strain energy is at least equal to the energy required to create the new crack surfaces. Let us consider a crack model as shown in the fig.6.33. A surface crack of length C is equivalent to an internal crack of length $2C$ of a plate with negligible width. A decrease in strain energy results forms the formation of a crack. Strain energy per unit plate thickness: $U_E = -\pi C^2 \sigma^2 / E$

where, σ is the tensile stress acting and C is half the length of internal crack. A negative sign is associated with the term because growth of the crack releases elastic strain energy.

The surface energy due to presence of the crack: $U_S = 4C\gamma_S^2$, where γ_S is surface energy per unit volume.

Change in energy due to presence of the crack: $\Delta U = U_E + U_S$ ----(1)

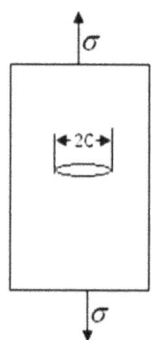

Fig.6.7. Griffth internal Crack Model

According to Griffith's criterion, the crack will propagate under constant stress if an incremental increase in the crack length produces no change in the total energy of the system. In other words the increased surface energy due to crack extension is compensated by the decrease in elastic strain energy.

Mathematically the condition is expressed as:

$$\frac{d(\Delta U)}{dC} = 0 \text{ or } \frac{d}{dc}(4C\gamma_S - \frac{\pi C^2 \sigma^2}{E}) = 0$$

or, $\frac{d}{dc}(4C\gamma_S - \frac{\pi C^2 \sigma^2}{E}) = 0$

or, $4\gamma_S = \frac{2\pi C^2 \sigma^2}{E}$

For $\sigma = \sigma_f$ we have: $\sigma_f = \sqrt{\left[\dfrac{2\gamma_S E}{\pi C}\right]} = \left[\dfrac{2\gamma_S E}{\pi C}\right]^{1/2}$ --------(2)

Where, σ_f is theoretical fracture strength according to Griffith's theory. Orowan suggested that the Griffith's equation for brittle fracture would be compatible with fracture in metals by inclusion of a term expressing the elastic work required to extend the crack wall. Accordingly, σ_f is revised as

$$\sigma_f = \left[\dfrac{2E(\gamma_S + \gamma_P)}{\pi C}\right]^{1/2} \quad --------(3)$$

Where, γ_P is the plastic defomation energy per unit volume which is in the order of to $10^2 - 10^3$ Joules per unit sq. meter surface area and γ_S is the surface energy of the crack per unit volume which is in the order of 1-2 Joules per sq. meter. Now:

$\sigma_f = \left[\dfrac{2E(\gamma_S + \gamma_P)}{\pi C}\right]^{1/2}$ can be approximated to

$\left[\dfrac{2E\gamma_P}{C}\right]^{1/2}$ as the value of γ_P far exceeds γ_S. Further the above equation was modified by Irwin with the term, G_c which is directly measurable as compared to the term γ_P. The modified equation becomes: $\sigma_f = \left(\dfrac{2EG_C}{\pi a}\right)^{1/2}$, where $G_C = \dfrac{\pi C \sigma_f^2}{E}$

which is the critical value of the Crack Extension Force or Strain Energy Release Rate. The critical value of crack extension force is also called the Fracture Toughness of the material.

****

Chapter 7
IMPACT TEST

7.1. Introduction:

Toughness of a material can be determined by calculating the area under the engineering stress-strain diagram. But such a calculation gives no indication regarding the behaviour of the material when subjected to sudden load. Further, the area remaining same under stress-strain diagram, the material may behave entirely in a different manner which not indicated by the stress-strain diagram obtained by a tensile test. Hence, an impact test is necessary to measure the relative toughness of the material along with its resistance to impact load. The impact test simulates the service conditions of the machine components which are often encountered in mining, transportation, agriculture & construction activities. All such equipments are frequently subjected to sudden loads. For example, when an automobile goes over a hump or falls into a ditch, the springs and axles experience a sudden rise in load. For avoiding failure this sudden loading must be resisted by them. A material which possesses higher impact resistance is said to be a tougher material. Hence, *toughness is the ability of a material to resists both fracture and deformation under impact loading.* Toughness is the parameter which includes both strength and ductility but is greatly different from both. For example, glasses & cast irons do have fairly good strength but lack in toughness as they fail easily under impact load. Similarly, highly ductile materials like gold, aluminium do not exhibit good toughness as they deform plastically quite easily. To be tough, a material must exhibit a good strength and should be fairly ductile to resist permanent deformation or failure.

7.2. Classification of Impact Tests:

There are two types of impact tests such as:

I. Charpy impact test and
II. Izod impact test.

7.3. Impact Testing Machine:

The ordinary impact testing machine has a swing pendulum of fixed weight as shown in the figure 7.1.

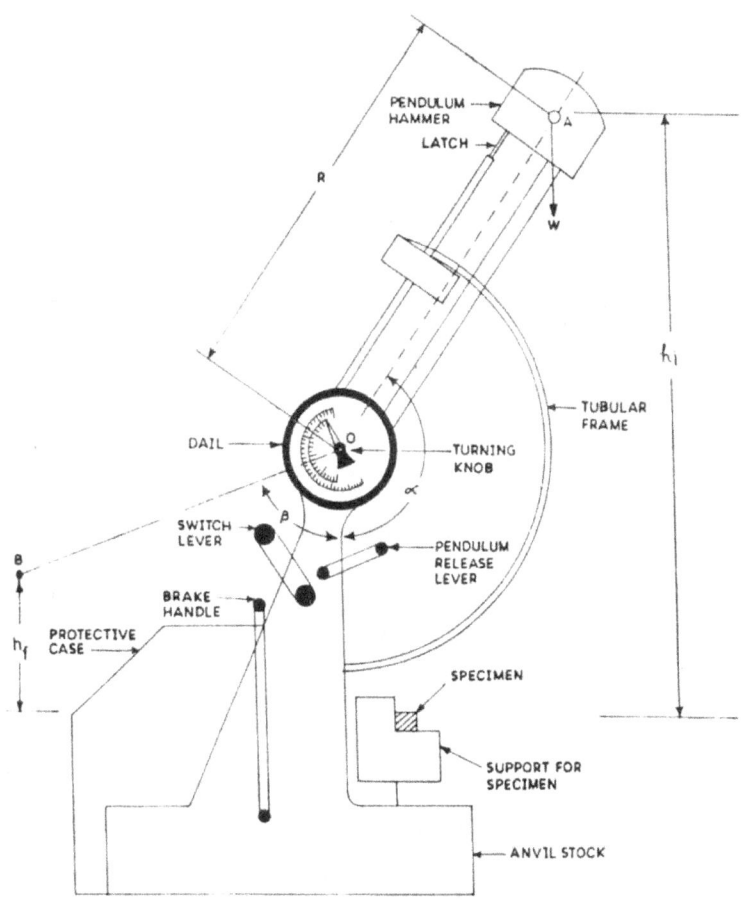

Fig.7.1. Details of an Impact Testing Machine

The pendulum is raised to a standard height depending on the type of specimen used. At the defined height, with reference to the clamping vice, the pendulum has a definite amount of potential energy expressed as *mgh*. This energy is indicated by a pointer-dial arrangement attached to the machine frame and pendulum. When the pendulum is released its potential energy gets converted to kinetic energy till it strikes the specimen. Certain amount of the kinetic energy gained by the pendulum is utilized in breaking the specimen and rest makes the pendulum to swing-up on the other side. Height of the swing-up on the other side is always less than the initial height form where the pendulum is released. Hence, the energy delivered to the specimen as impact load during testing is equal to the difference in potential energy of the pendulum at the initial height before the swing and the maximum potential energy of the pendulum at the end of the swing on the other side after striking the specimen. This energy difference is read from the dial-pointer arrangement which is (mgh_1-mgh_2), where: h_1 is the initial height, h_2 is the final height of the pendulum upto which it swings-up on the other side after breaking the specimen, *m* is the mass of the pendulum and *g* is the acceleration due to gravity(9.8m/sec^2). The impact toughness is measured in terms of energy i.e., joules consumed in breaking the specimen. Higher the energy consumed in breaking the mterial higher is the toughness.

7.4. Impact Test Specimen:

Generally V & U hole notched specimens are most widely used during impact tests. The specimen with the notches for both Charpy and Izod tests are indicated in the fig.7.2 with dimensions.

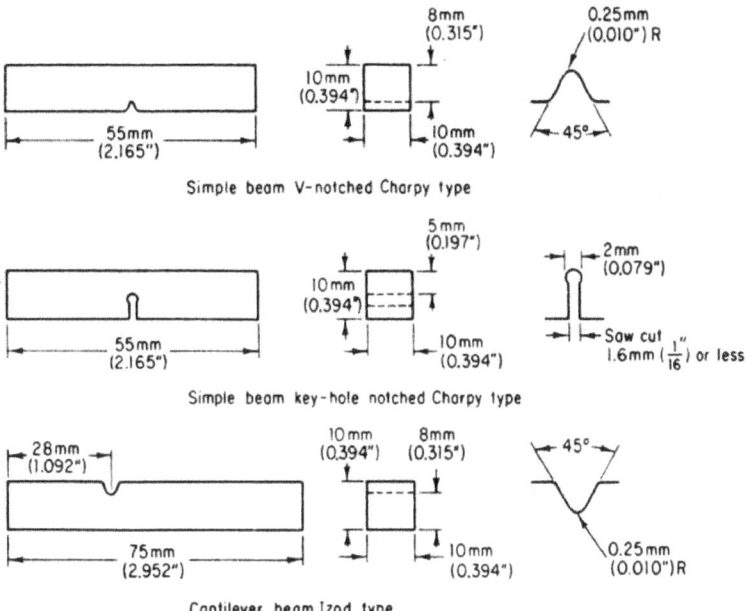

Fig. 7.2. Diagramatic view of different notches

i. The Charpy specimen is clamped in the vise so that it becomes a simple supported beam at both the ends.

ii. The Izod specimen is clamped in the vise, so that one of the ends of the specimen is free and it becomes a cantilever beam.

Method of loading in both the tests are different and is shown shematically in the fig.7.3.

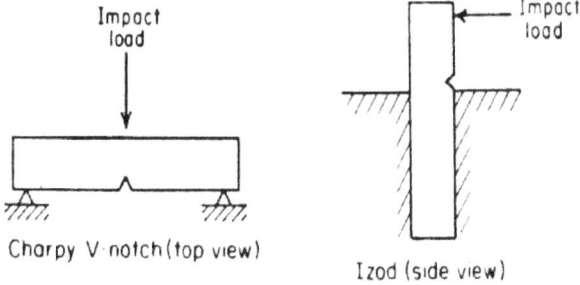

Fig. 7.3. Method of loading during Charpy and Izod impact tests

7.5. Informations obtained from the Impact Tests:

It is apparent that the notched-bar impact tests do not yield the true toughness of a material rather it reflects its behaviour with respect to a particular notch. However, the results are useful in a comparative sense. This is a routine test carried out in the aircraft and automotive industries as it has been found out that the high impact strength materials will provide satisfactory services of the components subjected to shock or impact load. Charpy test is widely conducted in USA while Izod test is widely used in Europe. However, Charpy test is accepted worldwide for determining the impact strength or toughness of the materials while the Izod test is mostly used for determining the impact resistance of the material.

I. Common information obtained from the impact test is the type of fracture that occurs in the material by examining the fractured surfaces. If the fractured surfaces appear fibrous, it is ascertained that the fracture is shear fracture and the material is relatively ductile. If the fractured surface appears granular, then the fracture is cleavage or brittle fracture. The fractured surface also may exhibit a dual nature. Usually an estimate is made regarding the percentage of area exhibited by the cleavage or brittle fracture. The area percentage estimated indicates the relative ductility or brittleness of the material.

II. Another important information gathered from the impact test is the energy consumed in fracturing the notched specimen. This value is only of relative importance and should not be used directly for any design purpose.

III. The notched bar impacts test is most useful when conducted over a range of temperature. The temperature at which ductile-to-brittle transition takes place can be determined by the impact test. This transition temperature is important in cryogenic application of the materials.

7.6. Relative Advantages of Charpy Test:

i. The principal advantage of Charpy notch test is that it is relatively simple, utilizes a small test specimen and quite low cost.

ii. This test can be carried out readily over a range of temperature or even at sub- ambient temperature.

iii. The design of the test specimen is well suited for measuring differences in notch toughness in low-strength materials such as structural steels.

iv. The test can be used to study the influence of alloying & heat treatment on notch toughness.

7.7. Relative Disadvantages of Charpy Test:

i. The major disadvantage of this test is that the test result cannot be used for design purpose as the measurement is not in terms of stress. It is extremely difficult to correlate impact energy data with service performance.

ii. The test result is not related to the size of the flaw present in the material.

iii. Large scatter in the transition temperature is inherent to the test which makes it difficult to plot a well defined transition temperature curve.

7.8. Transition Temperature:

Toughness of metallic materials strongly depends on temperature. Generally toughness of the metallic materials increases with increase in temperature as the plasticity increases. However, toughness of the metallic materials decreases rapidly with decrease in temperature. The notched bar impact test is most meaningful when conducted over a range of temperature as this test can determine the temperature at which ductile-to-brittle transition takes place.

The temperature at which the nature of the material changes from ductile to brittle is termed as *Brittle Transition Temperature or Transition Temperature*. The Impact energy C_v absorbed during failure versus Temperature curves are drawn to determine the trasition temperature as shown schematically in the fig 7.4.

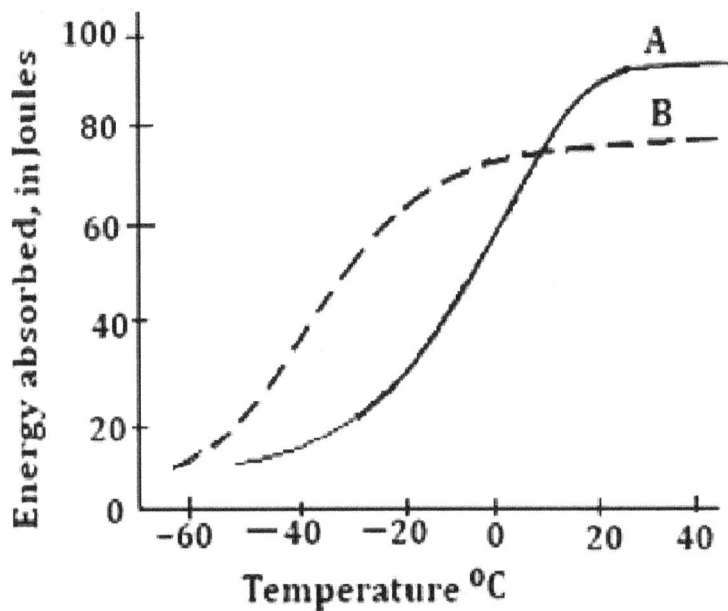

Fig.7.4. Effect of Teperature on Notch Toughness

By examining the curves it is observed that material *A* exhibits higher notch toughness at room temperature as compared to martial *B*. However, the brittle transition temperature of material *A* is much higher than that of material *B* which suggests that material *B* is better for applications at subzero temperatures than material *A* because of its lower brittle transition temperature.

7.9. Significance of Transition Temperature Curve:

Major utility of transition-temperature curves is to help in selecting materials which can resistant brittle fracture at the operating temperature. The brittle transition behaviour of a wide range of materials is classified into three distinct categories as shown in fig.7.5.

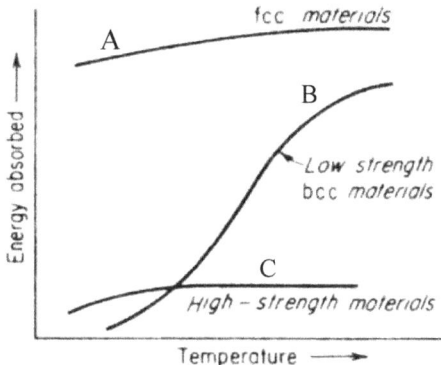

Fig.7.5. Brittle Trasition behaviour Diffent Materials

I. Curve A: It indicates medium and low strength FCC & HCP materials in which notch toughness is always higher. Brittle fracture is not a problem in these materials unless there is some special chemical environment.

II. Curve B: It indicates low & medium strength bcc materials like plain carbon steels, beryllium, zinc and ceramics demonstrating strongly temperature dependant brittle-transition temperatures. For such metals and alloys the brittle transition temperature is around 0.1-$0.2T_m$ while it is around 0.5-$0.6T_m$ for ceramic materials where T_m is the melting point of the material in Kelvin scale.

III. Curve C: It indicates high strength materials having tensile strength at least equal to $E/150$, brittle fracture occurs at a nominal stress around room temperature. These materials are high strength steels, Al & Ti alloys possessing low notch toughness.

7.10. Design Consideration of Transition Temperature:

The design should be such that the transition temperature is above the operating temperature so that brittle fracture will not occur within the elastic stress limit. To determine transition temperature, the typical C_v vs Temperature curve is drawn as shown schematically in the fig.7.6. As there is no single criterion to evaluate transition temperature various methods are usually employed to evaluate the same.

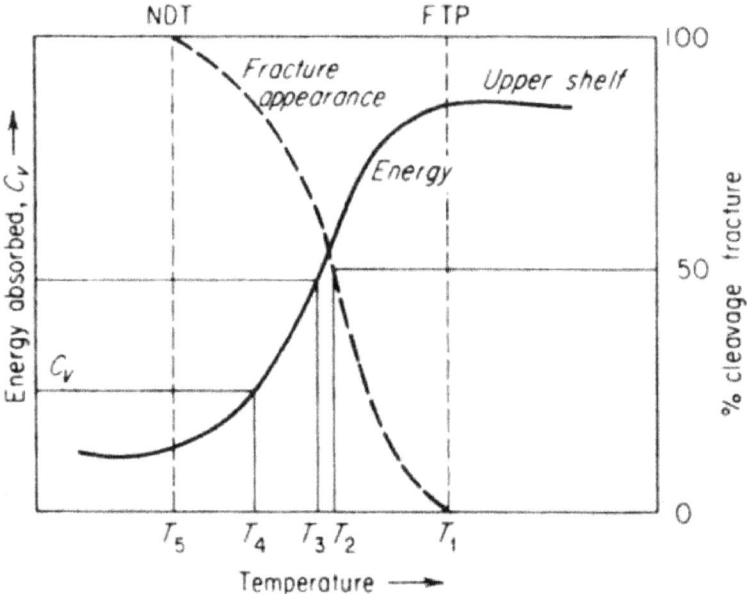

Fig.7.6. C_v vs Trasition Temperature Curve

7.11. Methods to Evaluate Transition Temperature:

I. The most conservative method is finding a temperature at where 100% fibrous fracture takes place. This transition temperature criterion is known as *Fracture Transition Plastic* (FTP). FTP is defined as the temperature where a total ductile fracture switch over to a completely brittle fracture.

II. A less conservative and more practical criterion is based on the transition temperature at which 50% cleavage (brittle) & 50% fibrous (ductile) fracture occurs. This is determined by studying the fractured surfaces at different temperature. This transition temperature is called *Fracture-Appearance-Transition Temperature* (FATT). The temperature T_2 shown in the fig.7.6. is the FATT.

III. A common criterion is to define transition temperature on the basis of an arbitrary low value of fracture energy C_v. This is often called *Ductile Transition Temperature* (DTT). By extensive trial tests it has been found out that brittle fracture is initiated around a C_v value of 20Joules. A 20J transition temperature (T_4) is significant for low strength steels only and is not significant for other materials.

IV. A well defined criterion is to base the transition temperature on the temperature at which the fracture becomes 100% cleavage (brittle). The temperature T_5 is called *Nil Ductility Temperature* (NDT) and this is the temperature at which fracture initiates without any prior plastic deformation. Further, below NDT the probability of ductile fracture is negligible.

7.11. Metallurgical Factors Affecting Transition Temperature:

1. Changes in transition temperature over 50°C can be achieved by changing the chemical composition or microstructure of low carbon steels. Largest change is obtained by changing the percentage of carbon and manganese. The 20J C_v transition temperature for steels is raised by about 14°C for each 1wt% increase in carbon in the steel. The effect of carbon on trasition temperature of plain carbon steel is shown schematically in fig.7.7.

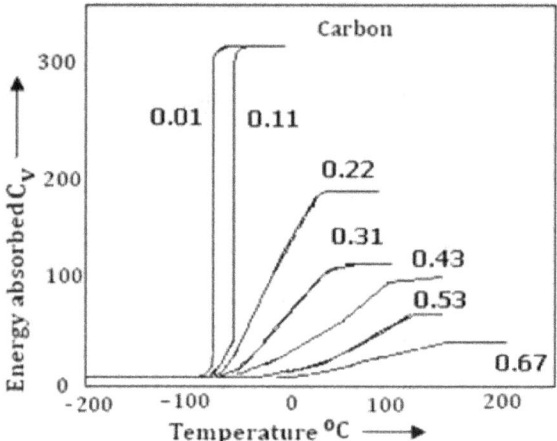

Fig.7.7. Effect of carbon on transition temperature of plain carbon steels

This transition temperature is lowered by 50°C for each 0.1wt% increase in the manganese content in the steels. The Mn:C ratio should be at least 3:1 for satisfactory notch toughness of plain carbon steels. The maximum practicable limit of Mn:C ratio is around 7:1. At this ratio the percentage of Mn in steel is around 1.4.

2. Phosphorus also has a strong effect on raising the transition temperature. 20J C_v transition temperature is raised by about 7°C for each 0.01wt% increase in phosphorus in the steel.

3. Nickel upto 2wt% in steel is beneficial as it raises the notch toughness. Molybdenum raises the transition temperatures as rapidly as carbon while chromium has little effect on transition temperature.

4. It is difficult to assess the effect of nitrogen on transition temperature in steels because of its interaction with other alloying elements. However, oxygen has a great effect on transition temperature of steels.

When oxygen content is raised from 0.001 to 0.06% the transition temperature increases from –150 to 340°C in steels. In view of this result it is not surprising that the deoxidizing practice of steel has a great effect on transition temperature. Rimmed steel with higher oxygen content shows a transition temperature of around 250°C. Semi-killed steels which are deoxidized with silicon have much lower transition temperature. The steels which are fully killed and deoxidized with silicon & aluminium exhibit 20J C_v transition temperature as low as – 60°C.

5. Grain size has a strong effect on the transition temperature. A decrease in grain size by half will increase the transition temperature by 160°C in mild steels.

6. Transition temperature of thick(≥ 12) hot rolled steel products is higher compared to thin plates with similar composition.

7. Transition temperature varies with the grain orientation of rolled or forged products.

****

Chapter 8

ERICHSEN CUPPING TEST

8.1. Introduction:

Erichsen cupping test is a ductility test which is carried out to measure the ability of a metallic sheet or strip to undergo plastic deformation by stretch forming. This test is essential to study the formability of a metal or alloy during press forming. The test consists of forming a cup by pressing the test piece with a spherical punch clamped between the blank holders, until through crack appears. The depth of the cup is measured just as the sheet metal starts to crack. This depth is called Erichsen Cupping Index (IE). Higher the cupping index higher is the ductility of the material under test. The Erichsen Cupping Test set-up is shown in the fig.8.1.

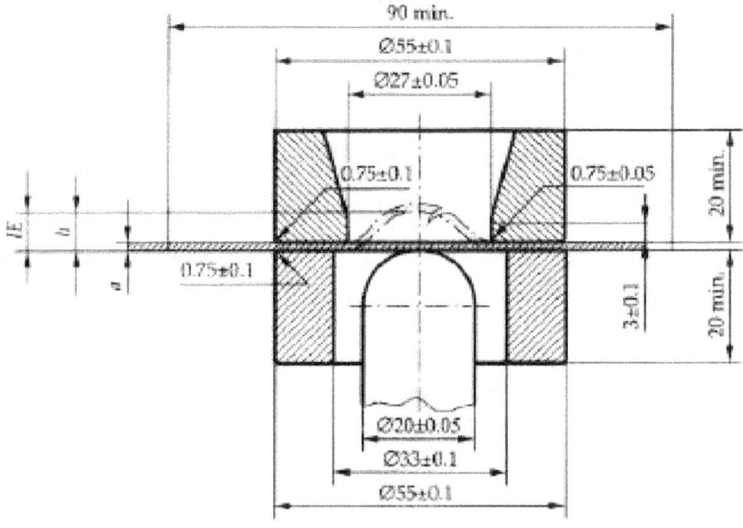

Key:

a = Thickness of the test piece
h = Depth of the indentation during the test
IE = Erichsen cupping index

Fig.8.1.Erichsen Cupping Test set-up

8.2. Test Procedure:

The Erichsen cupping test is carried on a machine equipped with a die, punch and blank holder. The machine is so constructed that it allows the observation of the bulging side of the test piece during the test. The loading is stopped instantly when through crack appears on the blank. The hardness of the punch and the die used in the cupping test should be around 750HV.

i. Measure the thickness of the specimen to the nearest 0.01mm.

ii. Lightly lubricate the surfaces of the specimen, punch & die with graphite grease.

iii. Clamp the specimen in the blank holder with a 10kN force.

iv. Bring the punch in contact with the specimen smoothly & start the measurement from this point.

v. Performe the test at a rate of 5-10 mm/ minute till crack appears.

vi. Terminate the movement of the punch at the instant when crack appears through the full thickness of the test piece. Measure the depth of penetration to the nearest 0.1mm. Conduct the test at least three times to calculate the mean depth of penetration or Erichsen Cupping Index in millimeters.

8.3. Test Conditions:

i. The specimen should be flat, smooth and free form foreign materials.

ii. There should be no burr on the edges of the specimen. The specimen may be rectangular or circular in shape.

iii. The width/diameter of the specimen should be at least 90 mm.

iv. The distance between the centre of any indentation & edge of the specimen should be at least 45mm and the distance between the centres of two adjacent indentations should be at least 90mm.

vi. Test should be carried out within 10-35°C.

******

Chapter 9

TORSION TEST

9.1. Introduction:

The ability of a material to resist twisting moment (torque) is determined by torsion test. This test has found wide acceptance similar to tension test. However, it is only useful in theoretical study of plastic deformation of metals and alloys. Torsion test is carried out on materials to determine:

i. Modulus of elasticity in shear.
ii. Torsional field strength.
iii. Modulus of rupture.

This test is highly essential for shafts, axles & twist drills and other such components which are subjected to torsion load as shown in the diagram 9.1.

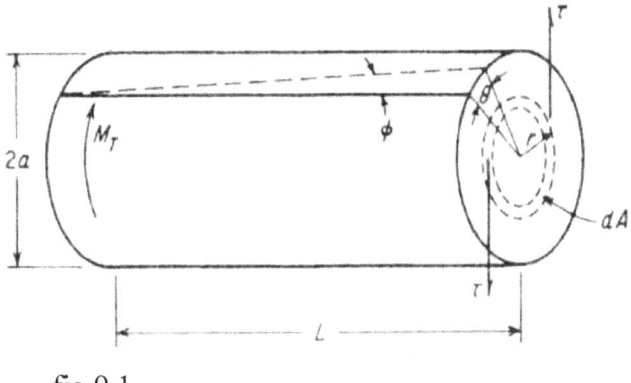

fig.9.1.

9.2. Torsion Test Equipment:

The testing equipment consists of a twisting head, with chuck for gripping and applying twisting moment to the specimen. The weighing head, which grips the other head of the specimen, measures the torque applied.

The angular displacement of a point on the specimen is measured with respect to a point on the same longitudinal element at the opposite end by a device called Troptometer. Torsion test is frequently carried out on circular or tabular specimen. Troptometer is used to determine the angle of twist λ n radians. If, L is the length of the specimen, the shear strain γ is given by equation:

$\gamma = \tan \lambda = \dfrac{r\lambda}{L}$, where, r is the radius of the section, λ is the angular displacement and L is the length of the specimen.

During the test, twisting moment M_T is measured and a torque-twist diagram is plotted.

9.3. Simple Torsion Test:

The simple torsion test is a ductility test employed to evaluate the ability of a metallic wire to undergo plastic deformation during simple torsion in one direction. One of the grips is capable of rotating while the other grip is fixed. The grips should also not prevent any contraction in length of the wire during testing to ensure a constant tensile load on the specimen.

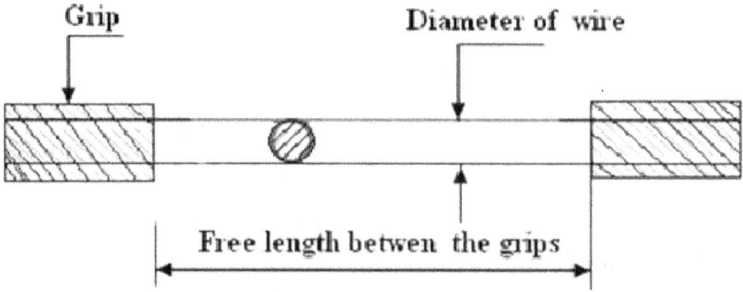

Fig.9.2. Princple of Simple Torsion test for Metallic Wire

****

Chapter 10
BEND TEST

10.1. Introduction:

The bend test is another ductility test employed to evaluate the ability of a metal to undergo plastic deformation in bending.

10.2. Specification of the Specimen:

Round, square, rectangular or polygonal test samples should be used for testing.

I. Width of the Specimen:

The width of the test piece should be as follows:

i. The test specimen should have the same width as that of the product under test if the product width is upto to 20mm.

ii. When the width of product is greater than 20mm with thickness less than 3mm the width of the specimen should be 20 ± 5mm.

ii. The test specimen should be 20-50 mm for products of thickness greater than equal to 3mm.

II. Thickness:

Specimen thickness should be same as that of product thickness when product thickness is less than 25mm. For products of thickness greater than 25mm, the thickness of the specimen may be reduced to 25mm by machining but should not below 25mm.

10.3. Different Types of Bend Tests:

i. Guided Bend Test:

This test ensures that the length of the support and the width of the mandrel are greater than the width or diameter of the test piece shown schematically in fig.10.1.

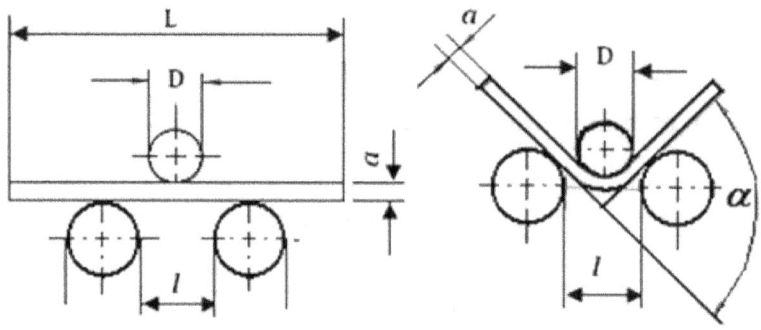

Key:
a = Diameter or thickness of the test piece
D = Diameter of the mandrel
L = Length of the test piece
l = Distace between the supports
α = Angle of the bend

Fig.10.1.Guided Three Point Bend Test

2. Semi-Guided Bend Test:

The specimen is placed on a V-block and force is applied continuously to achieve desired angle of bend or until fracture occurs.

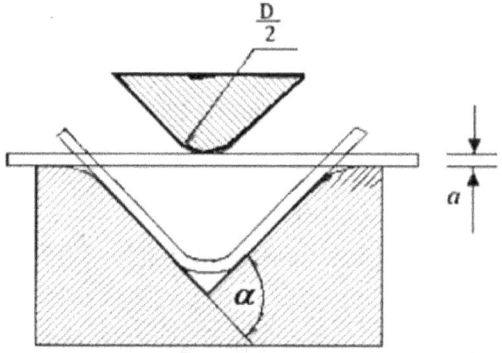

Key:
a = Diameter or thickness of the test piece
D = Diameter of the mandrel
α = Angle of the bend

Semi - guided – Bending device with V-block and mandrel

Fig.10.2. Semi-guided Bend Test

3. Free Bend Test:

The free bend test is carried out as shown in the figure below:

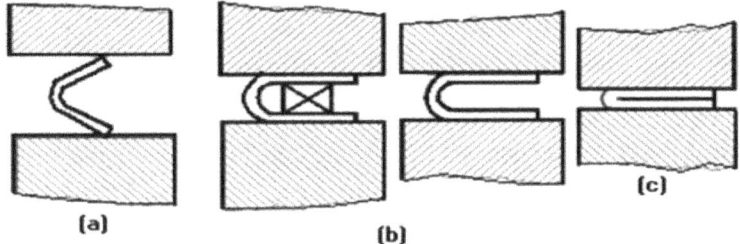

Fig.10.3. Free Bend Test

3.15. Applications of Bend test:

Modulus of rapture, σ_r for brittle materials may be determined by the bend test. The Load P is applied till the rupture of the beam takes place. Modulus of rapture σ_r can be approximated by the formula: $\sigma_r = \dfrac{MY}{I}$, where M is the maximum bending moment of a rectangular beam section, I is the area of moment of the beam's cross-section and Y is half depth of the beam. Accordingly, Modulus of rapture $\sigma_r = \dfrac{6M}{bh^2}$ and $\tau_r = \dfrac{3F}{2bh}$ where, F is the maximum shear force acting on the beam.

****

Chapter 11
NON DESTRUCTIVE TESTING OF MATERIALS

11.1. Introduction:

In the earlier chapters we have discussed extensively on testing methods to evaluate mechanical properties of materials. Most often the specimen which may or may not be a finished product is destroyed or made useless for future use during such mechanical tests performed on them. Hence, mechanical tests are known as Destructive Tests. It is also important to note that destructive test cannot be used on hundred percent of the finished components as it will destroy the components. The defects associated with the finished products may not be visible on a gross scale. For example, a small blow hole in a cast product may not hamper hardness or strength values but causes premature failue under service conditions. Further in many cases, if the defect is located at a particular place in the component due to some manufacturing error, it can be rectified if identified definitively without destroying the component. Further for all critical and non-critical use, most often hundred percent of the product is to be checked for surface smoothness, hardness and dimensional accuracy. But for checking many other properties, samples may have to be drawn from the finished product in a particular ratio. For more critical components like welded products & LPG cylinders, hundred percent of the weldments are to be checked and there can be no sampling in such cases. To check the quality of the weldments, we cannot adopt any destructive method as it will destroy the component itself. Once the component is in a finished state, its quality in terms of dimensional accuracy, hardness, internal soundness is to be fully ensured by certain test methods without causing any damage to the component.

So certain non-destructive testing techniques have been developed to ensure the quality of such components.

Nondestructive Testing:

A nondestructive test is an examination of an object in any manner which will not impair the future usefulness of the object. Although in most of the cases the NDT methods do not provide a direct measurement of mechanical properties, still they are extremely valuable in locating the defects in the material that would impair the performance of a machine member when placed in service. So NDT methods are used to detect defects in the components either before they are put into service or during their service life, permitting their removal before any failure takes place. Hence, NDT tests are used to make the products more reliable, safe and economical. It also helps in predicting the timely replacement of the machine components to avoid costly failures in future.

11.2. Elements of Non-Destructive Testing:

Five basic elements are required to conduct any nondestructive testing:

I. Source: A source which provides some probing medium namely a medium that can be used to inspect the component under test.

II. Modification: The probing medium must change or get modified due to the variations or discontinuities within the object under test.

III. Detection: A detector capable of determining the changes in the probing medium.

IV. Indication: A means of indicating or recording the signals from the detector.

V. Interpretation: A method of interpreting the indications observed or found.

The most common types of nondestructive testing or inspection are:

    i. Magnetic particle inspection.
    ii. Dye-penetrant inspection.
    iii. Ultra-sonic inspection.
    iv. Radiographic inspection.
    v. Eddy current inspection.

****

Chapter 12

MAGNETIC PARTICLE INSPECTION

12.1. Introduction:

Magnetic Particle Inspection (MPI) is a non-destructive testing method, which is used for detecting surface or near sub-surface discontinuities in ferromagnetic materials. The defects are usually fine cracks, laps, tears, seams, inclusions and similar such discontinuities which are extremely fine for the unaided eye but get magnified indirectly during testing.

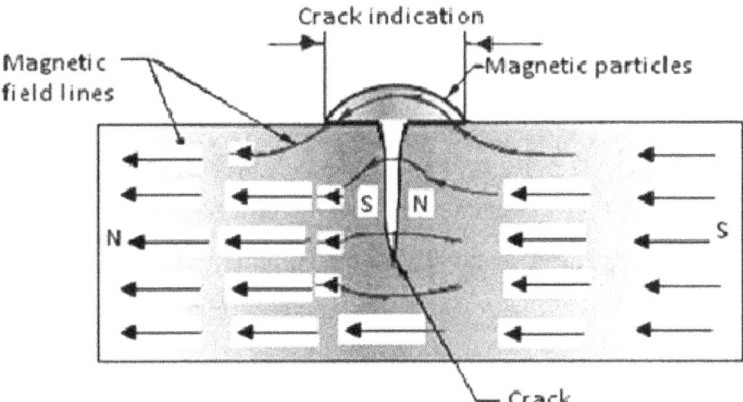

Fig.12.1. Principles of Magnetic Particle Inspection

12.2. Principle:

The study of magnetic lines of forces and their basic nature are extremely important to understand magnetic particle inspection.
i. Magnetic lines of forces are continuous non intersecting lines running from north pole to south pole.
2. The lines of force may deviate or terminate at any intermediate point for any localized change in the medium through which it passes.

The magnetic particle test consists of examining a part submitted to magnetic field coated with dry magnetic powder or with a liquid containing magnetic powder suspension. The surface or near subsurface discontinuity cause a distortion or deviation in the orientation of the magnetic lines of force around the defect giving a visible indication of the defect in the material.

12.3. Test Requirements:
i. Magnetizing Equipment:

The equipment intended to supply the magnetic flux.

ii. Demagnetization Equipment:

The equipment intended to reduce the residual magnetism in the component once the test is over.

iii. Detection Media:

Various detection media used during inspection are:

a. Dry magnetic power with coloured pigment.

b. Magnetic power suspended in aqueous or refined light petroleum oil medium.

c. Fluorescent magnetic power suspended in aqueous or well refined light petroleum oil medium.

12.4. Test Method:
i. The test piece should be demagnetized before inspection.

ii. The surface to be inspected should be free from scale, dirt, oil, paint and other contaminations. Non ferromagnetic coatings like paints upto $50\,\mu m$ thick do not normally impair detection sensitivity. Thicker coating reduces detection sensitivity.

iii. Magnetic field strength should be sufficiently high to produce satisfactory indications. However, it should not be too high to mask relevant indications.

iv. For proper flaw detection, the major axis of the discontinuity should be perpendicular to the magnetic lines of force.

v. The surface to be inspected should be adequately and uniformly coated with the magnetic particles. Coating may be applied by immersion, flooding, spraying or dusting.

There are two methods of magnetic inspection such as:

i. Residual Method:

In this method piece to be inspected should be magnetized initially and then covered with fine magnetic powder.

ii. Continuous Method:

In this method both magnetization and application of the inspecting medium are carried out simultaneously.

The magnetic particles applied on the work piece may be held as a suspension of dry or wet powder. The work piece under test is magnetized by two magnetic probes pressed onto the test piece from two sides. When the test piece is thoroughly uniform in the surface and sub-surface level, the magnetic particles are arranged uniformly along the magnetic lines of force from one pole to other without any discontinuity or intersection among them. However, when a surface discontinuity or defect is present, magnetic particle arrangement terminates around the defect. The change in uniform pattern is an indication of the defect being present and it assumes the approximate shape of the discontinuity.

Such a method of magnetic particle inspection is called Magnaflux as it was developed by Magnaflux Corporation.

Another variation of this test is Magnaglow where fluorescent magnetic particles are used in place of plain magnetic particles. In Magnaglow method, suspension is made to flow over the work piece. Fluorescent particles enter into the surface cracks by capillary action. Then the surface is viewed under black light (ultraviolet rays) which makes the cracks more visible. This method is mainly used to detect grinding, fatigue and casting cracks. Magnetic particle inspection finds extensive use in iron and steel foundries, forging and extrusion industries.

****

Chapter 13
ULTRASONIC FLAW DETECTION

13.1. Introduction:

Ultrasonic flaw detection method locates and measures the size of the internal defects present in the materials which cannot be inspected by the magnetic particle or dye penetrant method. This test method uses a high frequency sound wave, ordinarily known as ultrasonic waves to locate the internal defects in the materials.

If a piece of metallic material is struck by a hammer, it produces audible notes. The pitch and damping of such audible notes(sound waves) produced is strongly influenced by the presence of internal voids, flaws or defects in the material. It gives an indication of the continuity or discontinuity (soundness) of the material. This method helps only in determining large internal defects. Large casting or forgings are selected by listening to the sound they produce when struck. Same principle is adopted for selecting coconuts.

13.2. Principle of Ultrasonic Flaw Dection:

This method utilizes a very large frequency sound wave in the range of 1-5 million hertz to detect internal flaws in the material. Such high frequency ultrasonic sound waves are usually produced by piezo-electric crystals. Sound waves are simply organized mechanical vibrations traveling through a medium, which may be a solid, a liquid, or a gas. These waves will travel through a given medium at a specific speed or velocity, in a predictable direction, and when they encounter a boundary with a different medium they will be reflected or transmitted according to simple rules. High frequency ultrasonic sound waves have the ability to penetrate the metals and liquids of large thickness.

13.3. Types of Ultrasonic Flaw Detection:

Ultrasonic flaw detection is quite simple and classified into two basic types:
1. Through-Transmission
2. Pulse-Echo

1. Through -Transmission Ultrasonic Flaw Detection:

In this method the transmitted beam is analyzed which has undergone natural decay in intensity which passing through the medium. The extent of natural decay in intensity during passage through a medium depends on the physical properties of the medium itself and they are correlated. However, if there is a discontinuity in the path of the waves, large portion of the wave energy is reflected back from the defect and the intensity of the sound wave reaching the receiver side decreases considerably. This considerable decrease in the intensity of the received signals indicates a defect or flaw in the material. This test method is called Through Transmission and is shown schematically in the fig.13.1.

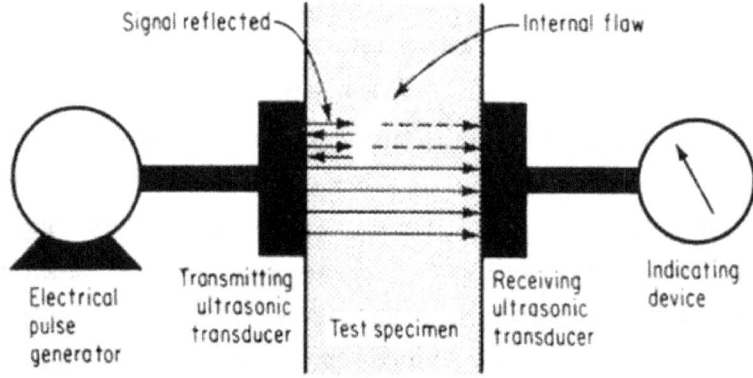

Fig.13.1.Through -Transmission Ultrasonic Flaw Detection:

2. Pulse-Echo Method:

In the pulse-echo method only one transducer is used which serves as both transmitter and receiver. When the material under test is completely uniform, part of the pulse transmitted from the transducer returns or reflects back into the material from the back surface. This reflection from the back surface which is the echo is seen as a peak in the wave form in the oscilloscope. The scheme is shown in the fig.13.2.

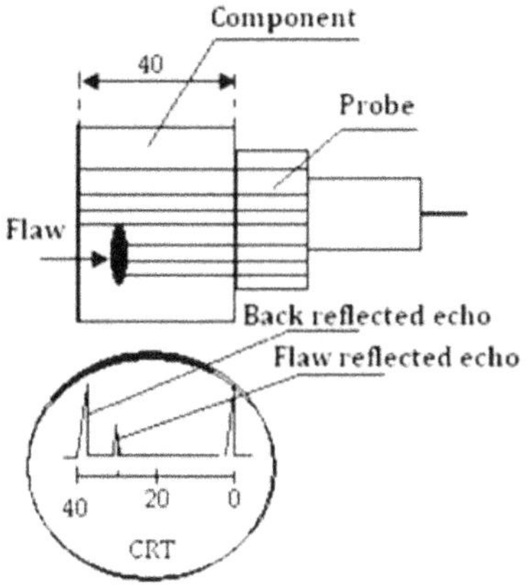

Fig.13.2. Pulse-Echo Method:

Hence, for a totally defect free material, only two voltage peaks are observed on the oscilloscope screen in the pulse-echo method, one corresponds to the initial pulse from the transmitter and the other one being the reflected pulse form the back surface of the material. When the material contains a flaw, another voltage peak is observed in between the initial and final peaks. This happens due to the fact that, a portion of the transmitted ultrasonic beam gets reflected by the flaw and returns back to the receiving transducer and results in another voltage peak on the oscilloscope screen at some intermediate position.

Since the peaks on the oscilloscope screen measures the elapsed time between the reflection of the pulse from the front and back surfaces, the distance between the peaks can be taken as a measure of the material thickness. Therefore, the location of a defect may accurately be determined form the positions additional peaks on the oscilloscope screen. Higher the frequency, smaller the defects that can be detected. Ultrasonic inspections are adopted to detect and locate defects like shrinkage, cavities, internal bursts, cracks, porosity and large nonmetallic inclusions in castings. Wall thickness can also be measured in closed vessels.

13.4. Precautions taken during Ultrasonic Flaw Detection:
i. Orientation of the flaw should be proper with respect to the transmitted pulse. The test piece should be rotated in order to ensure that the flaw becomes perpendicular to the transmitted beam for obtaining high intensity echo of the flaw on the oscilloscope screen. This helps in locating the flaw in the material.
ii. Probe and work piece contact should be as good as possible. As ultrasonic test is usually carried out on finished products as a quality check, proper contact between the probe & work piece is extremely important for the success of this test. Usually oil and grease are used to established proper contact between probe and work piece.
iii. If small parts are to be tested, they are to be place inside oil, water or glycerin tank for ensuring proper contact. Care is to be taken for proper transmission of waves in the test piece on which the reliability of the test depends.

13.5. Advantages of Ultrasonic Flaw Dection:

i. High penetrating power, which allows detection of deep flaws.

ii. High sensitivity, permitting detection of extremely small flaws.

iii. In many cases only one surface needs to be accessible.

iv. Greater accuracy than other non-destructive methods in determining the depth of internal flaws and the thickness of parts with parallel surfaces.

v. Some capability of estimating the size, orientation, shape and nature of defects.

vi. Some capability of estimating the structure of alloys of components with different acoustic properties.

vii. Non hazardous to operations or to nearby personnel and has no effect on equipment and materials in the vicinity.

viii. Capable of portable or highly automated operation.

ix. Results are immediate. Hence on the spot decisions can be made.

13.6. Disadvantages of Ultrasonic Flaw Dection:

i. Manual operation requires careful attention by experienced technicians.

ii. Extensive technical knowledge is required for the development of inspection procedures.

iii. Parts that are rough, irregular in shape, very small or thin, or not homogeneous are difficult to inspect.

iv. Surface must be prepared by cleaning and removing loose scale, paint, etc., although paint that is properly bonded to a surface need not be removed.

v. Couplants are needed to provide effective transfer of ultrasonic wave energy between transducers and parts being inspected unless a non-contact technique is used.

vi. Inspected items must be water resistant, when using water based couplants that do not contain rust inhibitors. In these cases anti-freeze liquids with inhibitors are often used.

****

Chapter 14
EDDY CURRENT TESTING

14.1. Introduction:

Eddy current technique is employed to inspect electrically conducting materials for their surface defects, irregularities, variation in composition at the surface level and electroplating thickness on the base materials. Eddy current (EC) testing is a no contact method for the inspection of metallic parts.

1.4.2. Principle:

When a metallic component is placed inside an alternating current (a.c.) coil, eddy current is induced into the surface of the metallic components. The magnitude and frequency of the induced eddy current is related to the following factors:

i. Magnitude and frequency of the alternating current in the coil.

ii. Electrical conductivity of the specimen.

iii. Magnetic permeability of the specimen.

iv. Shape of the specimen.

v. Relative position of the coil and specimen.

vi. Metallurgical nature of the specimen.

The induced eddy current is generally concentrated near the surface of the resulting in so called "*skin effect*". Mathematically, the depth of penetration of the eddy current: $D = 1/(\pi f \mu \sigma)^{1/2}$, where D is the depth of penetration of eddy current in meters, f is the frequency of the a.c. source, μ is the Magnetic permeability of the material (Henry per meter) and σ is the Volumetric electrical conductivity of the material (Mhos per meter).

From the above mathematical relation, it can be seen that the depth of penetration D increases with decrease in frequency f of the alternating current and vice versa. For detecting surface and near surface flaws a higher test frequency is used while for probing deep into the surface a lower test frequency a.c. is used.

14.3. Use of Eddy Current Testing:

i. For detecting longitudinal & transverse cracks, inclusion in the tubular or cylindrical section like, pipes, rods, tubes &etc. The most important application of eddy current testing is the continuous testing of extruded products like pipes, rods, tubes, bars & etc.

ii. Another important industrial use of this testing is sorting of components for heat treatment variation or compositional mix up. This application requires use of two eddy current coils. A standard test specimen is kept in one of the coils, while the components to be tested are passed through the second coil. Acceptance or rejection of the components under test is determined by comparing the waveforms from both on an oscilloscope screen. If the test piece waveform matches with that of the standard test piece it is accepted or else rejected.

iii. It can also be used on a continuous or non continuous basis for measuring coating or electroplating thicknesses.

iv. A larger area can be scanned in a single-probe pass, while maintaining a high resolution.

v. Less need for complex robotics to move the probe; a simple manual scan is often enough.

vi. Complex shapes can be inspected using probes customized to the profile of the part being inspected.

****

Chapter 15

X-RAY RADIOGRAPHY

15.1. Introduction:

X-rays are high energy short range electromagnetic radiations. It has the ability to penetrate into opaque objects. This specific property of the X-rays is exploited to assess flaws within the materials. Similar to X-rays Gamma rays can be used for the purpose. Gamma rays have more penetrating power compared to X-rays but less sensitive. Due to lack of sensitivity gamma-rays have found limited applications compared to X-rays.

15.2. Production of X-Rays:

X-rays are produced when a metallic material known as target material is bombarded with high velocity steam of electrons. When the high velocity electrons are stopped by the target material, a part of the energy possessed by the electrons is converted to radiation energy or X-rays. An X-ray unit is shown schematically in the fig.15.1.

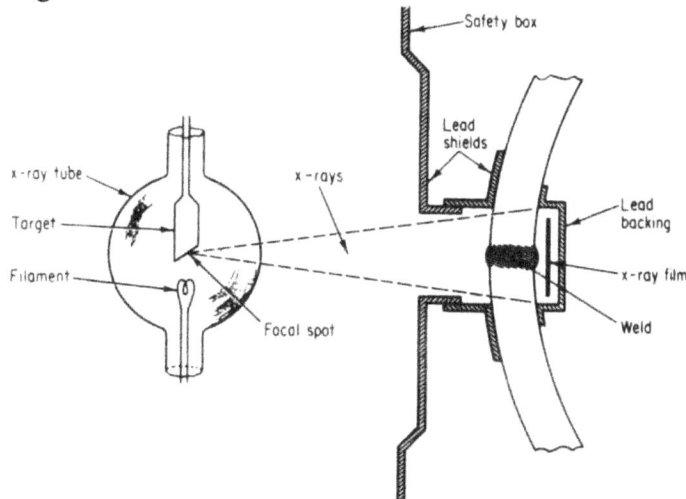

Fig.15.1.Schematic X-ray Unit

15.3. Requirements for X-Rays Generation:

1. Electron Source:

A filament (cathode) to provide stream of electrons towards the target material.

2. Target Material:

A target material, usually copper anode is located in the path of electrons.

3. Regulate Power Supply:

A high voltage difference is maintained between the cathode and anode. Thus high velocity electrons are produced and are made to strike the target material. The the voltage difference between the cathode and anode regulates the wave length of X-rays produced.

4. Current Controller:

There is also a means required to regulate the tube current to control the number of electrons striking the target.

15.4. Physical Principles of X-ray Radiography:

1. Wave length of the X-ray produced:

The Shortest wavelength of the X-ray produced is:
$$\lambda_{swl} = 12.40 \times 10^3 / V$$

2. Absorption:

When X-rays fall on solid materials which are opaque to visible light beam, two distinct phenomena take place:

i. A portion of the incident beam gets absorbed &

ii. The other portion gets transmitted through the solid material. Experiments have shown that the initial intensity I_o of the X-ray beam, as it passes through any homogeneous medium, gets reduced. The fractional decrease in intensity is proportional to the distance 'x' traversed in the medium.

The reduced intensity I_x is related to the incident beam, I_o with a mathematical relation: $I_x = I_o e^{-\mu x}$, where, I_o is the intensity of the incident X-ray beam, I_x is the intensity of the transmitted X-ray beam, μ is the absorption coefficient of X-rays, considered to be an intrinsic property of the materials and x is the distance traversed by the X-ray through the medium or material. Hence, the reduced intensity I_x of the X-ray depends on the mass and the nature of the material through it travels. This basic nature of absorption intensity can be utilized to locate and study sub-surface flaws in the materials. When the X-ray radiograph is carried out on a welded or cast product containing some flaws like gas holes, cracks, blowholes, sand or slag inclusions, the absorption pattern at the flaw location will be different. The transmitted beam is recoded on a photographic plate and the place on the photographic film corresponding to the area of discontinuity or flaw will be different compared to the rest of the area. So, the flaw can be identified easily by examining the photographic plate as shown in the fig15.2.

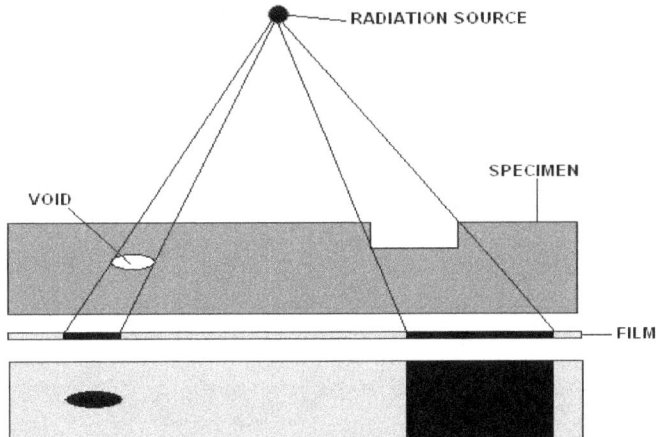

Fig15.2. Plan view of the X-ray film where darker areas represent defects

15.5. Major Applications of X-ray Radiography:

i. X-ray radiography has found intensive application for evaluating soundness of weldments in pressure vessels, high pressure pipelines, LPG cylinders and other critical components where cent percent checking & quality assurance is imperative.

ii. X-ray can also be used to measure the thickness of a material. Schematic view of the thickness guage is a shown in the figure.15.3. Initially the thickness guage is standardized for a particular thickness of the material. If the thickness changes the amount of transmitted radiations reaching the detector also changes and the detector shows different reading. This thickness guage can be used for online checking of the flat rolled products in the industries.

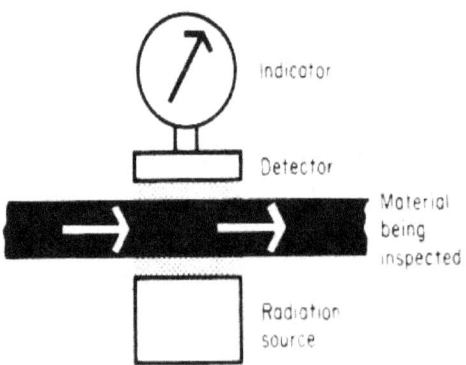

Fig.15.3. Thickness Guage

****

Chapter 16

DYE PENETRANT INSPECTION

16.1. Introduction:

Dye-penetrant inspection is also known as Liquid or Fluorescent penetrant inspection (LPI or FPI). It is a nondestructive testing method used for detecting surface discontinuity such as cracks, seams, laps, cold shuts, laminations, isolated porosity & shrinkage open to the surface. This method is universal in nature as it can be applied well to ferrous, nonferrous, plastics and ceramics.

16.2. The Test - Sequence of Operations:

i. Preparation of the surface to be inspected.

ii. Application of the liquid penetrant to the prepared surface.

iii. Removal of the excess penetrant.

iv. Application of the developer and.

v. Visual examination and assessment.

16.2.1. Surface Preparation and Application of Penetrant:

The surface to be inspected is made clean by the cleaner. Then the surface is treated with the dye penetrant by dipping, spraying or by dusting. The penetrant is an oil like liquid which is seeps into the fissures or cracks by capillary action.

16.2.2. Removal of the Excess Penetrant:

After the penetrant is drawn into the cracks the remaining penetrant on the surface is washed off. Removal of extra penetrant ensures penetrant only in the cracks or fissures.

16.2.3. Developer Application to locate Crack:

Then the part is treated with the developer which is may be dry powder, suspension of powder in water or suspension of powder in non-aqueous volatile solvents. The developer powder acts like a sponge and draws back the penetrant from inside the cracks & defects. Penetrant on reaction with the developer produces some significant colouration on the surface which marks out the location of the defect with simultaneously enlargement of its size. To make the observation more clear & simple, contrasting colours for the penetrant and developer are used. A combination white developer with red penetrant is most commonly used. The sequence of operations is shown schematically in fig.16.1.

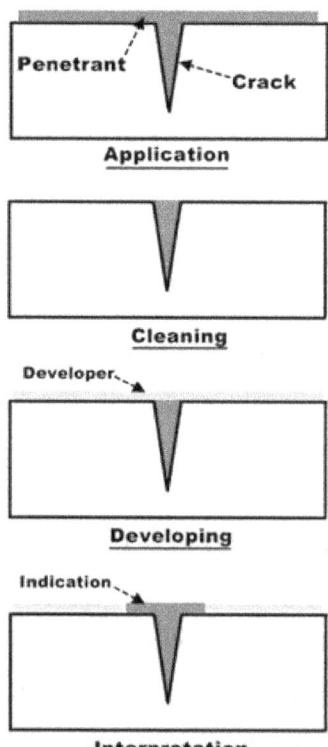

Fig.16.1. Essential Operations of Liquid/Dye Penetrant Inspection.

16.3. Variation in Dye-Penetrant Inspection Method:

A little variation of dye penetrant inspection is fluorescent penetrant method. The steps involved here are exactly identical to that of dye penetrant method excepting the only difference of using fluorescent material in the penetrating liquid. When the defective components are inspected or viewed under ultraviolet light (black light) the defects appear as bright fluorescent marks against the dark background. Fluorescent or dye-penetrant inspection method is used extensively to locate cracks and shrinkage in castings, fabricated components, welded structures paticularly in case of steam or gas turbine blades.

****

Chapter 17

GLOSSARY OF TERMS

Abrasion: The process of rubbing, grinding or wearing by friction.

Alclad: Composite sheet produced by bonding either corrosion-resistant aluminium alloy or aluminium of high purity to base metal of structurally stronger aluminum alloy.

Allotropy: The reversible phenomenon by which certain metals may exist in more than one crystal structure. If not reversible, the phenomenon is termed "polymorphism".

Alloy: A substance having metallic properties and being composed of two or more chemical elements of which at least one is a metal.

Amorphous: Not having a crystal structure: noncrystalline.

Black light: Electromagnetic radiation not visible to the human eye. The portion of the spectrum generally used in fluorescent inspection falls in the ultraviolet region of 3300-4000A°, with the peak at 3600A°.

Blue brittleness: Brittleness exhibited by some steels after being heated to some temperature within the range of 150 to 350^{0}C and more especially if the steel is worked at the elevated temperature.

Brittle crack propagation: A very sudden propagation of a crack with the absorption of no energy except that is stored elastically in the body. Microscopic examination may reveal some deformation even though it is not noticeable to the unaided eye.

Brittle fracture: Fracture with little or no plastic deformation.

Brittleness: Quality of a material that leads to crack propagation without appreciable plastic deformation.

Carbide tools: Cutting tools, made of tungsten carbide, titanium carbide, tantalum carbide, or combinations of them, in a matrix of cobalt or nickel, having sufficient wear resistance and heat resistance to permit high speed machining.

Charpy test: A pendulum type single blow impact test in which the specimen, usually notched, is supported at both ends as a simple beam and broken by a falling pendulum. The energy absorbed as determined by the subsequent rise of the pendulum, is a measure of impact strength or notch toughness.

Cohesion: Force of attraction between the molecules (or atoms) within a single phase. Contrast with adhesion.

Cohesive strength: (1) The hypothetical stress in an unnotched bar causing tensile fracture without plastic deformation. (2) The stress corresponding to the forces between atoms.

Cooling stresses: Residual stress resulting from non uniform distribution of temperature during cooling.

Corrosion embrittlement: The severe loss of ductility of metal resulting from corrosive attack, usually intergranular and often not visually apparent.

Corrosion fatigue: Effect of the application of repeated or fluctuating stresses in a corrosive environment characterized by shorter life than would be encountered as a result of either the repeated or fluctuating stresses alone or the corrosive environment alone.

Creep: Time dependent strain occurring under stress. The creep strain occurring at a diminishing rate is called primary creep; that occurring at a minimum and almost constant rate, secondary creep; that occurring at an accelerating rate, tertiary creep.

Creep strength: (1) the constant nominal stress that will cause a specified quantity of creep in a given time at constant temperature, (2) The constant nominal stress that will cause a specified creep rate at constant temperature.

Critical strain: The strain just sufficient to cause the growth of very large grains during heating where no phase transformations take place.

Crystal: A solid composed of atoms, ions, or molecules arranged in a pattern which is repetitive in three dimension.

Crystalline Fracture: A fracture of a polycrystalline metal characterized by a grainy appearance.

Crystallization: The separation of a solid crystalline phase usually from a liquid phase on cooling.

Cubic plane: A plane perpendicular to any one of three crystallographic axes of the cubic (isometric) system, the miller indices are (100).

Cup and Cone fracture: Fracture, frequently seen in tensile test piece of a ductile material, in which one of portions of the surface of failure shows a central flat area of failure in tension with an exterior extended rim of failure in shear.

Defect: A condition that impairs the usefulness of an object.

Deformation bands: Parts of a crystal which have rotated differently during deformation to produce bands of varied orientation within individual grains.

Diamond-pyramid hardness test: An indentation hardness test employing a diamond pyramid indenter and variable loads enabling the use of one hardness scale for all range of hardness from very soft lead to tungsten carbide.

Dilatometer: An instrument for measuring the expansion or contraction in a metal resulting from changes in such factor as temperature or allotropy.

Dislocation: A linear defect in the structure of a crystal. Two basic types are recognized, but combinations and partial dislocation are most prevalent. An "edge dislocation" corresponds to the row of mismatched atoms along a straight edge formed by an extra, partial plane of atoms within the body of the crystal that is by a plane of smaller area than any other parallel section through the crystal. A "screw dislocation" corresponds to the highly distorted lattice adjacent to the axis of a spiral structure in a crystal, the spiral structure being characterized by a distortion that has joined normally parallel planes together to form a continuous helical ramp winding about the dislocation as an axis with a pitch of one interplaner distance.

Distortion: Any deviation from the desired shape or contour.

Drawabiliity: A measure of the workability of a metal subject to a drawing process. A term usually expressed to indicate a metal's ability to be deep-drawn.

Drawing: (i) Forming recessed pats by forcing the plastic flow of metal in dies. (ii) Reducing the cross section of wire or tubing by pulling it through a die.

Ductile crack propagation: Slow crack propagation that is accompanied by noticeable plastic deformation and requires energy to be supplied from outside the body.

Ductility: The ability of material to deform plastically without fracturing, being measured by elongation or reduction of area in a tensile test.

Dye penetrant: Penetrant with dye added to make it more readily visible under normal lighting condition.

Eddy-current testing: Nondestructive testing method in which eddy-current flow is induced in the test object. Changes in the flow caused by variations in the object are reflected into a nearby coil(s) of subsequent analysis by suitable instrumentation and techniques.

Elastic deformation: Change of dimension accompanying the stress in the elastic range, original dimensions being restored upon release of stress.

Elasticity: That property of a material by virtue of which it tends to recover its original size and shape after deformation.

Elastic limit: The maximum stress to which a material may be subjected without any permanent strain remaining upon complete release of stress.

Elongation: In tensile testing, the increase in the gage length, measure after fracture of the specimen within the gage length, usually expressed as a percentage of the original gage length.

Extensometer: An instrument for measuring changes caused by stress in a liner dimension of a body.

Fatigue: The phenomenon leading to fracture under repeated or fluctuating stresses having a maximum value less than the tensile strength of the material. Fatigue fractures are progressive, beginning as minute cracks that grow under the action of the fluctuating stress.

Fatigue life: The number of cycles of stress that can be sustained prior to failure for a state test condition.

Fatigue limit: The maximum stress below which a material can presumable endure an infinite number of stress cycles. If the stress is not completely reversed, the value of mean stress, the minimum stress or the stress ratio should be stated.

Fatigue strength: The maximum stress that can be sustained for a specified number of cycles without failure, the stress being completely reversed within each cycle unless otherwise stated.

Fibrous fracture: A fracture where the surface is characterized by a s dull gray or silky appearance. Contrast with crystalline fracture.

Fibrous structure: (i) In forgings, a structure revealed as laminations, not necessarily detrimental, on an etched section or as a ropy appearance on a fracture. It is not to be confused with the "silky "or "ductile" fracture of a clean metal. (ii) In wrought iron, a structure consisting of slag fibers embedded in ferrite.

File hardness: Hardness as determined by the use of a file of standardized hardness on the assumption that a material which cannot be cut with the file is as hard as, or harder than the file. Files covering a range of hardness may be employed.

Fluorescent magnetic-particle inspection: Inspection with either dry magnetic particle or those in a liquid suspension, the particle being coated with a fluorescent substance to increase the visibility of the indication.

Fluoroscopy: An inspection procedure in which the radiographic image of the subject is viewed on a fluorescent screen, normally limited to low-density materials or thin sections of metal because of the low light output of the fluorescent screen.

Formability: The relative ease with which a metal can be shaped through plastic deformation.

Fractography: Descriptive treatment of fracture, especially in metals with specific reference to photographs of the facture surface. Macrofractography, involves photographs at low magnification; macrofractography at high magnification.

Fracture stress: (i) the maximum principal true stress at fracture. Usually refers to unnotched tensile specimens. (ii) The (hypothetical) true stress which will cause fracture without further deformation at any given strain.

Fracture test: Braking specimen and examining the fractured surface with the unaided eye or with a low-power microscope to determines such things as composition, grain size, case depth, soundness or presence of defects.

Gauge length: The original length of that portion of the specimen over which strain, change of length and other characteristics are measured.

Granular fracture: A type of irregular surface produced when metal is broken; characterized by a rough grain-like appearance as differentiated from a smooth and silky or a fibrous type. It can be sub classified into transgranular and intergranular forms. This type of fracture is frequently called crystalline fracture.

Grinding cracks: Shallow cracks formed in the surface of relatively hard material because of excessive grinding heat or the high sensitivity of the material.

Hardness: Resistance of metal to plastic deformation usually by indentation. However the term may also refer to stiffness or temper or to resistance to scratching, abrasion or cutting. Indentation hardness may be measured by various harness tests such as Brinell, Rockwell and micro hardness.

Hot–shortness: Brittleness in metal in the hot-forming range.

Impact energy (impact value): The amount of energy required to fracture a material, usually measured by means of an Izod or Charpy test. The type of specimen and testing conditions affect the values and therefore should be specified.

Impact test: A test to determining the behaviour of materials when subjected to high hate of loading usually in bending, tension or torsion. The quantity measured is the energy absorbed in breaking the specimen by a single blow, as in the Charpy or Izod test.

Indentation hardness: The resistance of a material to indentation. This is the usual type of hardness test, in which a pointed or rounded indenter is pressed into a surface under a substantially static load.

Intracrystalline: Between the crystals or grains of a metal.

Intracrystalline: Within or across the crystals or grains of a metal. Same as transcrystalline and transgranular.

Isotropy: Quality of having identical properties in all directions.

Izod test: A pendulum type of single-blow impact test in which the specimen, usually notched is fixed at one end and broken by a falling pendulum. The energy absorbed as measured by this subsequent rise of the pendulum is a measure of impact strength or notch toughness.

Knoop hardness: Microhardness determined from the resistance of metal to indentation by a pyramidal diamond indenter having edge angle of 172 30′ and 130 making a rhombohedral impression with one long and one short diagonal. This long diagonal is measured microscopically to determine the KHN.

Lead screen: In radiography, a screen used
(1) to filter out soft-wave of scattered radiation and
(2) to increase the intensity of the remaining radiation so that the exposure time can be decreases.

Luder bands: Surface markings or depression resulting from localized plastic deformation in metals which show discontinuous yielding.

Magnetic-particle inspection: A nondestructive method of inspection for determining the existence and extent of possible defects in ferromagnetic materials, finely divided magnetic particles, applied to the magnetized part, are attracted to and outline the pattern of any magnetic-leakage fields created by discontinuities.

Malleability: The characteristic of metal which permits plastics deformation in compression without rupture.

Matrix: The principal phase or aggregate in which another constituent is embedded.

Mechanical properties: The properties of material that reveal its elastic and inelastic behaviour where force is applied, there by indicating its suitability for mechanical application; for example modulus of elasticity, tensile strength, elongation, hardness and fatigue limit.

Mechanical twin: A twin formed in metal during plastic deformation by a simple shear of the lattice

Miller indices (plane): Indices which identify a family of planes in crystal structure. The intercepts m, n, and p for any plane within a crystal give the reciprocals *1/m, 1/n,* and *1/p*, which may be changed to a common denominator, resulting in the number *h, k* and *l*, respectively. These numerators when written as (*hkl*) identify the family of planes to which the specific plane belongs.

Modulus of elasticity: Measure of the stiffness of metal. It is the ratio of stress, within proportional limit to the corresponding strain.

Nondestructive inspection: Inspection methods that do not destroy the part to determine its suitability for use.

Octahedral plane: In cubic crystals a plane with equal intercepts on all three axes for example (111) and any such plane.

Offset: The distance along the strain coordinate between the initial portion of a stress strain curve and a parallel line that intersects the stress-strain curve at a value of stress which is used as a measure of the yield strength. It is used for materials that have no obvious yield point. A value of 0.2 percent is commonly used.

Optical pyrometer: An instrument for measuring the temperature of heated material by comparing the intensity of light emitted with a known intensity of an incandescent lamp filament. **Any** device for measuring temperatures above the range of liquid thermometers.

Penetrant inspection: A method of nondestructive testing for determining the existence and extent of discontinuities that are open to the surface in the part being inspected. The indications are made visible through the use of a dye or fluorescent chemical in the liquid employed as the inspection medium.

Permanent set: Plastic deformation that remains upon releasing the stress that produced the deformation.

Plasticity: The ability of a metal to deform nonelastically without rupture.

Polymorphism: The ability of a material to exist in different crystal forms at different temperatures which is irreversible in nature is generally termed as Polymerphism.

Proportional limit: The maximum stress at which strain remains directly proportional to the stress.

Radiograph: A photographic shadow image resulting from uneven absorption of radiation in the object being subjected to penetrating radiation.

Radiography: A nondestructive method of internal examination in which metal or other objects are exposed to a beam of x-ray or gamma radiation. Difference in thickness density or absorption caused by internal discontinuities are apparent in the shadow image either on a fluorescent screen or on photographic film placed behind the object.

Recovery: Reduction or removal of work-hardening effects, without motion of large-angle grain boundaries.

Recrystallization: (i) the change from one crystal structure to another, as occurs on heating or cooling through a critical temperature. (ii) The formation of a new, strain-free grain structure form that existing in cold-worked metal, usually accomplished by heating.

Recrystallization annealing: Annealing cold-worked metal to produce a new grain structure without phase change.

Recrystallization temperature: The approximate minimum temperature at which complete recrystallization of a highly cold-worked metal occurs within a specified time, usually one hour.

Reduction of area: (i) Commonly the difference expressed as a percentage of original area between the original cross-sectional area of a tensile test specimen and the minimum cross-sectional area measured after complete separation.
(ii) The difference, expressed as a percentage of original area, between original cross-sectional area and that after straining the specimen.

Residual method: Method of magnetic-particle inspection in which the particles are applied after the magnetizing force has been removed.

Residual stress: Stress present in a body that is free of external forces or thermal gradients.

Resilience: (i) the amount of energy per unit volume release upon unloading. (ii) the capacity of a metal, by virtue high yield strength and low elastic modulus to exhibit considerable elastic recovery upon release of load.

Rockwell hardness test: A test for determining the hardness of a material based upon the depth of penetration of a specified penetrator into the specimen under certain arbitrarily fixed conditions of test.

Scleroscope test: A hardness test where the loss in kinetic energy of a falling diamond tipped metal "tup", absorbed by indentation upon impact of the tup on the metal being tested is indicated by the height of rebound

Scratch hardness: The hardness a metal determined by the width of a scratch made by a cutting point drawn across the surface under a given pressure.

Shear: (i) That type of force which causes or tends to cause two contiguous part of the same body to slide relative to each other in a direction parallel to their plane of contact. (ii) A type of cutting tool with which a material in the form of wire, sheet, plate or rod is cut between two opposing blades.

Shear angle: The angle that the shear plane in metal cutting, makes with the work surface.

Shear fracture: A fracture in which a crystal (or a polycrystalline mass) has separated by sliding or tearing under the action of hear stresses.

Shear strength: The stress required to produce fracture in the plane of cross section the conditions of loading being such that the directions of force and of resistance are parallel and opposite although their paths are offset a specified minimum amount.

Shortness: A form of brittleness in metal. It is designated as cold, hot and red to indicate the temperature range in which the brittleness occurs.

Shot Peening: Cold working the surface of a metal by metal-shot impingement.

Silly fracture: A metal fracture in which the broken metal surface has a fine texture usually dull in appearance. It is the characteristic of tough and strong metals.

Slip band: A group of parallel slip line so closely spaced as to appear as a single line when observed under an optical microscope.

Slip direction: The crystallographic direction in which the translation or slip takes place.

Slip line: The trace of the slip plane on the viewing surface; the trace is (usually) observable only if the surface has been polished before deformation. The usual observations on metal crystals (under the light microscope) are of a cluster of slip lines.

Slip plane: The crystallographic plane in which slip occurs in a crystal.

S-N diagram: A plot showing the relationship of stress S and the number of cycle N before failure in fatigue testing.

Stress-rupture test: A tension test performed at constant load and constant temperature, the load being held at such a level as to cause rupture. It is also known as creep - rupture test.

Stretcher strains: Elongated marking that appears on the surface of some material when deformed just at the yield point. These markings lie approximately parallel to the direction of maximum shear stress and are the result of localized yielding. Same as Luder lines.

Stringer: In wrought material an elongated configuration of micro-constituent or foreign material aligned in the direction of working. Commonly the term is associated with elongated oxide or sulfide inclusion n steel.

Superficial Rockwell hardness test: Form of Rockwell hardness test using relatively light loads which produce minimum penetration. Used for determining surface hardness or hardness of thin sections or small parts or where a large hardness impression might be harmful.

Superlattice: an ordered arrangement of atoms in a solid solution to form a lattice super imposed on the normal solid-solution lattice.

Tensile strength: In tensile testing, the ratios of maximum load to original cross sectional area. It is also called ultimate strength.

Thermal analysis: A method for determining transformation in a metal by noting the temperature at which thermal arrests occur. These arrests are manifested by change in slope of the plot or mechanically traced heating and cooling curves. When such data are secured nearly equilibrium conditions of heating and cooling, the method is commonly used for determine in certain critical temperature require for the construction of equilibrium diagram.

Thermal fatigue: Fracture resulting from the presence of temperature gradients which vary with time in such manner as to produce cyclic stress in a structure.

Thermocouple: A device for measuring temperature consisting of two dissimilar metals which produce an electromotive force roughly proportional to the temperature difference between their hot and cold junction ends.

Torsion: A twisting action resulting in shear stresses and strains.

Toughness: Ability of a metal to absorb energy and deform plastically before fracturing. It is usually measured by the energy absorbed in a notch-impact test, but the area under the stress-strain curve in tensile testing is also a measure of toughness.

Twin band: On a polished and etched surface, the section through a twin and the parent crystal.

Ultimate strength: The maximum conventional stress-tensile, compressive or shear-that a material can withstand.

Vickers hardness: Microhardness determined from, the resistance, of a metal to indentation by a 136° Diamond-Pyramid indenter making a square impression.

Yield point: The first stress in a material usually less than the maximum attainable stress at which an increase in strain occurs without an increase in stress. Only certain meals exhibit a yield point. If there is a decrease in stress after yielding a distinction may be made between upper and lower yield points.

Yield strength: The stress at which a material exhibits a specified deviation from proportionally of stress and strain. An offset of 0.2% is used for many metals.

****

BIBLIOGRAPHY

1. Mechanical Metallurgy: GE Dieter, McGraw Hill 1988
2. Material Science & Engineering: JD Callister, Jr, John Wiley & Sons
3. Structure & Properties of Materials, Vol-III: J Wullf, John Wiley
4. Physical Metallurgy Principles: R E Reed-Hill, East-West Press
5. Introduction to Physical Metallurgy: SH Avner, TMH
6. Manufacturing Technology: P N Rao, TMH